Heathkit
Educational Systems

CONCEPTS OF ELECTRONICS

Book 2

Copyright © 1981
Seventeenth printing—1986
Heath Company
Not Affiliated with D.C. Heath Inc.
All Rights Reserved
Printed in the United States of America

Model EB-6140
HEATH COMPANY
BENTON HARBOR, MICHIGAN 49022
595-2585   ISBN 0-87119-065-6

Unit 3

ACTIVE DEVICES

CONTENTS

INTRODUCTION

Basically, there are two types of components used in electronic circuits, passive and active devices. **Passive** devices do not alter their resistance, impedance or reactance when constant AC signals are applied to them. Examples of passive devices are resistors, inductors, and capacitors. **Active** devices, however, change their resistance or impedance when varying voltages are applied to them. As a result of this, active devices can amplify and rectify AC signals. Examples of active devices are vacuum tubes, transistors, and diodes.

Let's take a brief look at the development of active devices. In the 1920's, the vacuum tube revolutionized the field of electronics, allowing the invention of many devices including the radio receiver. By 1950, solid-state diodes and transistors, which are smaller and more rugged, started to replace vacuum tubes. The early 1960's saw the construction of groups of transistors and diodes on a single chip of silicon, forming an integrated circuit. The result of this development has been more compact and reliable electronic circuits and equipment. You will be studying several of these solid-state devices in this unit.

The unit objectives listed on the next page state exactly what you are expected to learn from this unit. Study this list now and refer to it often as you study the text.

UNIT OBJECTIVES

When you complete this unit, you should be able to:

1. State the difference between P-type and N-type semiconductor materials.

2. Define depletion region.

3. Identify both forward and reverse-biased diodes.

4. Name the two most important solid-state diode ratings and define each one.

5. State the characteristics of zener and varactor diodes.

6. Identify the schematic symbols of a solid-state diode, a zener diode, and a varactor diode.

7. Identify the schematic symbols for NPN and PNP transistors.

8. Name the three sections of a transistor and identify them on a transistor schematic symbol.

9. State the correct bias for the emitter-base and collector-base junctions of a transistor.

10. Identify and state the characteristics of the common-emitter, common-base, and common-collector amplifier configurations.

11. Define beta, thermal runaway, and maximum power dissipation as applicable to transistors.

12. Identify the schematic symbols of a JFET, a depletion mode MOSFET, and an enhancement mode MOSFET and name the terminals.

13. State the difference between depletion and enhancement mode MOSFETs.

14. Name the three basic FET circuit configurations and state the characteristics of each circuit.

15. Define light, infrared rays, ultraviolet rays and photon.

16. State the light spectrum's frequency range.

17. Name four light sensitive devices, state the characteristics of each, and identify their schematic symbols.

18. State the basic operating principles of the light emitting diode (LED) and identify its schematic symbol.

19. Determine the necessary value of bias resistor for correct LED operation.

20. State the operating characteristics and modes of the liquid crystal display.

21. Define integrated circuit and list its advantages and disadvantages.

22. Name the two basic types of integrated circuits.

23. State the basic types and characteristics of digital ICs.

24. State the basic characteristics of linear ICs and the operational amplifier.

25. Find amplifier voltage gain when given input and output voltage.

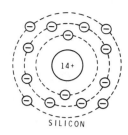

Figure 3-1

The atomic structure of silicon and germanium.

SOLID-STATE DIODES

Solid-state or semiconductor diodes have been in use since the earliest days of radio. However, recent advances have improved the semiconductor diode, and specialized diodes such as the zener diode and variable capacitance diode have been developed. In this section, you will learn the basic principles of semiconductors. You will also study the solid-state diode.

Semiconductor Materials

In an earlier unit, a semiconductor was described as being neither a good conductor nor a good insulator. This is due to the atomic structure or more importantly, the valence or outer shell electrons. Figure 3-1 shows the atomic structure of the two semiconductors used in most solid-state devices: silicon and germanium. Note that both have 4 valence electrons. Compare this to the best conductors, which have one valence electron, or to the best insulators which have a full valence shell of eight electrons.

Silicon and germanium, in their pure form, are actually worse conductors than is indicated by their four valence electrons. This is due to their crystalline structure. A simplified diagram of the germanium crystal structure is shown in Figure 3-2. Note that only the valence electrons are shown. In the crystal structure, each of the four valence electrons of any one atom is shared with four neighboring atoms. Thus, each atom appears to have eight electrons in its valence shell. This fills each atom's valence shell and it becomes very difficult to free an electron for current flow.

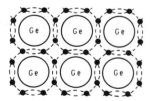

Figure 3-2
Simplified diagram of germanium crystal structure.

By adding carefully controlled quantities of certain impurities to,the pure semiconductor, its conductivity can be increased. This is called **doping**. An example of an impurity atom is arsenic, which has 5 valence electrons. When it is added to a semiconductor crystal, only 4 of its 5 valence electrons can be fitted into the crystal structure. The additional fifth electron becomes free to act as a current carrier. This is shown in Figure 3-3. A crystal doped in this way is known as **N-Type** semiconductor material, since it contains additional electrons.

Another impurity atom used to dope semiconductors is gallium. However, it has only three electrons. Therefore, when it is added to a semiconductor crystal a deficiency of electrons occurs. As shown in Figure 3-4, there now exists an area or **hole** in the crystal structure that lacks an electron. These holes behave as positively charged particles which are free to drift throughout the crystal. Due to the presence of the holes, the doped material is known as a **P-Type** semiconductor.

While a hole is not actually positive, it is an area that a randomly drifting electron can "fall" into, thus completing the crystal structure. However, once an electron "falls" into a hole, another hole is created in the region from which the electron came. The movement of a hole in this way is equivalent to the movement of a positive charge equal to the negative charge carried by one electron.

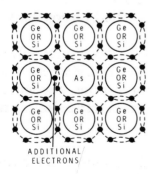

Figure 3-3
Semiconductor material doped with arsenic (N-Type).

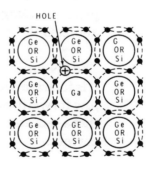

Figure 3-4
Semiconductor material doped with gallium (P-Type).

P-N Junctions

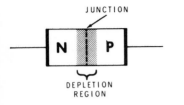

A

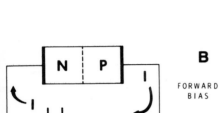

B

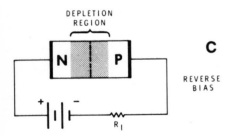

C

Figure 3-5
Solid-state diode operation.

When N and P type semiconductor materials are grown together to form a single crystal, a **solid-state diode** results. This is shown in Figure 3-5A. The area where the P and N type materials join is called the junction. As soon as the junction is formed, there will be a movement of electrons across it. Electrons near the junction will move from the N-type material into the P-type and fill the holes. As a result of this, the N-type material near the junction is depleted of electrons, while the P-type material near the junction is depleted of holes. Therefore, this area is called the **depletion region**.

Figure 3-5B shows what happens when a battery is connected to the PN junction. In this case, a negative voltage is applied to the N-type material, while a positive voltage is applied to the P-type. In this condition, once the voltage is above 1/2 to 1 volt, the applied voltage forces the electrons to cross the depletion region and continue across the P-type material. In other words, current flows through the diode and the depletion region no longer exists. This condition is called **forward bias**. Note that resistor R_1 must be added to limit the current. This is because the voltage drop across the junction is very low.

Figure 3-5C shows what happens when the battery connections are reversed. There is no current flow and the depletion region becomes larger. This is because the negative voltage on the P-type material forces electrons into the holes, thus depleting the region of still more holes. The positive voltage on the N-type material attracts the free electrons, which depletes the region of more electrons. This is called **reverse-bias**.

The PN-junction acts as a one way switch because it conducts only when it is forward biased. It is called a **solid-state diode**. And, since it has a very low forward voltage drop, it makes an excellent rectifier. You will recall that a rectifier is used to convert alternating current into pulsating direct current.

The schematic symbol for the solid-state diode is shown in Figure 3-6. The N-type material is called the **cathode**, while the P-type is the **anode**. The arrowhead in the schematic symbol points in the direction **opposite** to current flow. Note that when the solid-state diode is forward biased, current flows from cathode to anode.

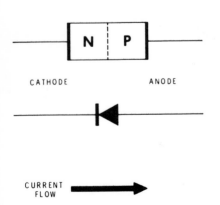

Figure 3-6
Schematic symbol for a solid-state diode.

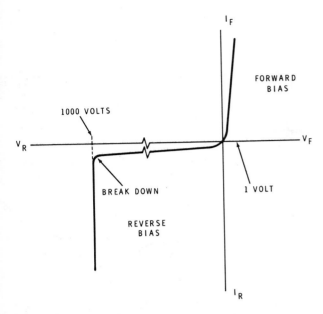

Figure 3-7
Diode characteristic curve.

Diode Ratings

Figure 3-7 shows the "characteristic curve" for a solid state diode. This curve shows diode current under forward and reverse bias conditions. In the forward bias condition, it shows that the forward voltage drop (V_F) is a relatively constant 1 volt regardless of the forward current (I_F).

An important diode rating is the maximum forward current. If this rating is exceeded, the diode will be destroyed due to excessive heat. A typical maximum forward current rating is 1 A, however, some diodes may be rated as high as 20 A or as low as 10 mA.

The curve of Figure 3-7 also shows reverse bias characteristics. Note the small amount of reverse current (I_R). Since this current is normally a few microamperes, it can usually be ignored. However, when the reverse voltage (V_R) reaches 1,000 V, the reverse current drastically increases. This is due to a **breakdown** of the depletion region. The electrons are "ripped" from the valence shell and a high current results. Normally, this will damage the diode; however, some diodes are specially designed to operate in breakdown. They are called zener diodes and are discussed in the next section.

Since a breakdown of the diode during reverse bias is possible, diodes have a **peak inverse voltage** (PIV) rating. The PIV is the maximum reverse bias voltage the diode can handle before breakdown. Typical PIV ratings are 1,000 to 2,000 V. However, diodes can be rated as high as 7,000 V or as low as 30 V.

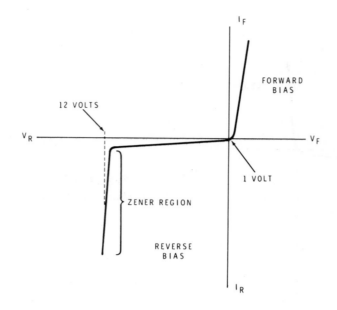

Figure 3-8
Zener diode characteristic curve.

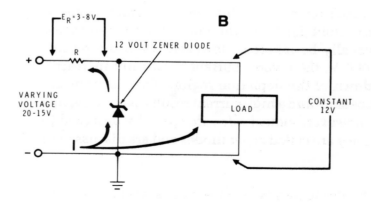

Figure 3-9
The zener diode's (A) schematic symbol, and
(B) voltage regulator circuit.

Zener Diode

Figure 3-8 shows the characteristic curve of a typical zener diode. Notice that just after breakdown, in this case at 12 V, the voltage (V_R) is constant for any value of reverse current (I_R). This is called the zener region, and once the diode is biased into this region, its voltage drop is constant. While all diodes have this characteristic, zener diodes are specially designed to dissipate the heat generated when operated in breakdown. They are also designed to break down at standard voltage values. The zener diode's schematic symbol is shown in Figure 3-9A.

The constant voltage drop across a zener diode can be put to good use by placing a load in parallel with it. Figure 3-9B shows a circuit which takes a varying voltage of 15-20 V and, using a zener diode, presents a constant 12 V to the load. This circuit is known as a voltage regulator because it supplies a constant voltage to the load, regardless of input variations. Voltage regulators are used to supply voltage to circuits which are sensitive to voltage variations.

Varactor Diode

Another special purpose diode is the varactor or variable capacitance diode. This diode uses the variable width of the depletion region in reverse bias. Figure 3-10A shows that, for varying voltages, the width of the depletion region varies proportionately. Since the depletion region acts as an insulator, it becomes the dielectric and the P and N type material become the plates of a capacitor. Now, by varying the reverse bias voltage and the plate spacing, the diode capacitance is also varied. While all diodes exhibit some junction capacitance, varactors are designed to enhance this property.

Applications of the varactor include using it as part of a resonant circuit. Then the resonant frequency can be changed by varying the diode's voltage. The schematic symbol for a varactor is shown in Figure 3-10B.

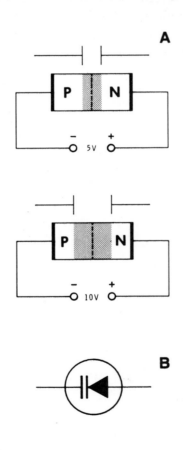

Figure 3-10
Varactor diode (A) operation,
(B) schematic symbol.

Self-Review Questions

1. A doped semiconductor material which contains additional electrons is called a/an _____ _____ semiconductor. A material which has holes or a deficiency of electrons is called a/an _____ semiconductor.

2. In a solid-state diode, the area near the PN junction which has a deficiency of electrons and holes is called the _____ _____ .

3. In Figure 3-11A, the diode is _____ biased, while in
   _{forward/reverse}
   Figure 3-11B the diode is _____ biased.
   _{forward/reverse}

4. Name the two most important solid-state diode ratings and define each one. _____
_____
_____

A

B

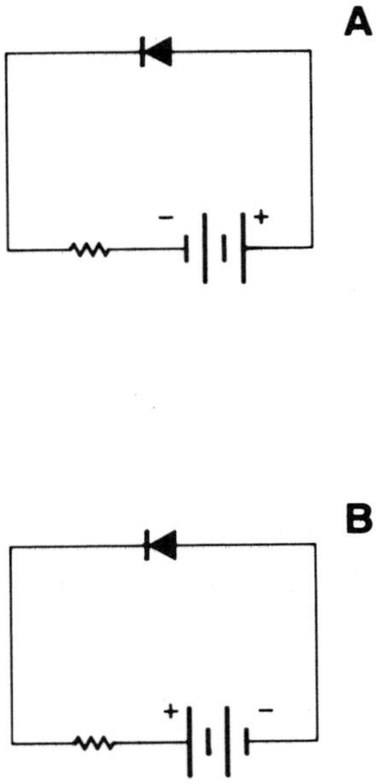

Figure 3-11
Are these diodes forward or reverse biased?

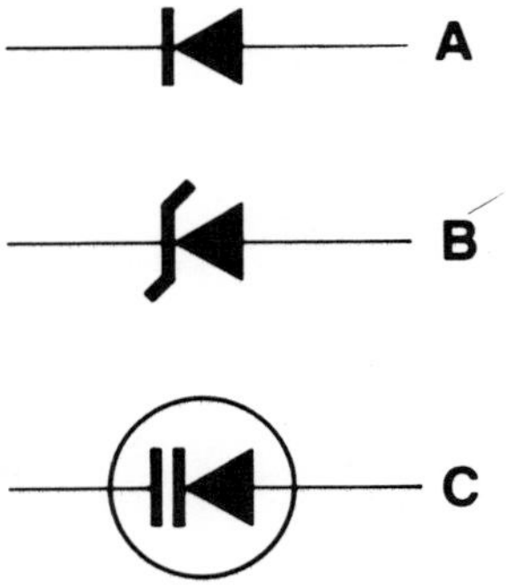

5. Identify the schematic symbols shown in Figure 3-12.

   A. _____

   B. _____

   C. _____

6. A zener diode is operated in _____ bias and has a
   _{forward/reverse}
   _____ .

7. A varactor diode is operated in _____ bias and has a
   _{forward/reverse}
   voltage variable _____ .

A

B

C

Figure 3-12
Identify these schematic symbols.

Self-Review Answers

1. A doped semiconductor material which contains additional electrons is called an **N-type** semiconductor. A material which has a deficiency of electrons or holes is called a **P-type** semiconductor.

2. In a solid-state diode, the area near the PN junction which has a deficiency of electrons and holes is called the **depletion region.**

3. In Figure 3-11A, the diode is **forward** biased, while in Figure 3-11B the diode is **reverse** biased.

4. **PIV** or **peak inverse voltage** is the maximum amount of reverse bias voltage a diode can withstand before breakdown occurs.

   Maximum Forward Current is the maximum forward bias current that a diode can handle.

5. The schematic symbols shown in Figure 3-12 are:

   A. **Solid-state diode**

   B. **Zener diode**

   C. **Varactor diode**

6. A zener diode is operated in **reverse** bias and has a **constant voltage drop**.

7. A varactor diode is operated in **reverse** bias and has a voltage variable **capacitance**.

TRANSISTORS

An extension of the solid-state diode is the transistor. However, the transistor uses two PN-junctions, rather than one, and can actually **amplify** input signals. Amplification occurs when the output signal is greater than the input signal. Therefore, the transistor has many more applications and is much more versatile than the diode.

In this section, you will be learning basic transistor operation. You will also be introduced to the three common transistor amplifier configurations.

Transistor Structure

A transistor is a single semiconductor crystal that contains two PN junctions. It is constructed so that a P-type section is "sandwiched" between two N-type sections or an N-type section is "sandwiched" between two P-type sections. Thus, there are two transistors; NPN and PNP. Figure 3-13 shows both types of transistor structure and their schematic symbols.

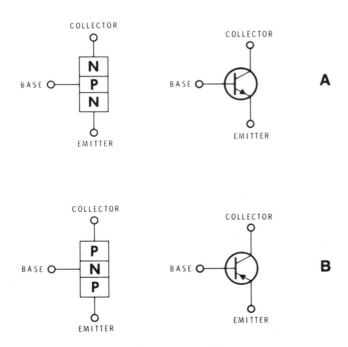

Figure 3-13
Transistor structure and schematic symbols.
(A) NPN transistor,
(B) PNP transistor.

In both NPN and PNP transistors, the three sections are called the **emitter**, the **base**, and the **collector**. In the schematic symbol, the emitter is identified by an arrowhead. Just as in the diode, the arrowhead points in the direction **opposite** to electron flow. It also points toward the N-type material. This will help you distinguish between an NPN and PNP schematic symbol.

Transistor Operation

The operation of a transistor junction is similar to the operation of a diode junction. However, the transistor is used to amplify, while the diode can only rectify. The diode is normally used as a one-way switch. Therefore, it is changing between forward and reverse bias. In a transistor, however, the emitter-base junction is forward biased and the collector-base junction is reverse biased. Figure 3-14 shows correctly biased PNP and NPN transistors.

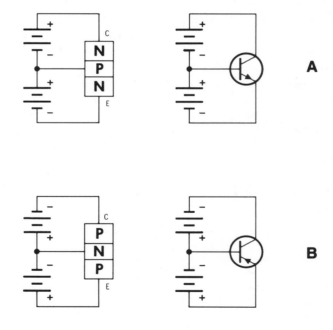

Figure 3-14
Correctly biased (A) NPN transistor, and
(B) PNP transistor.

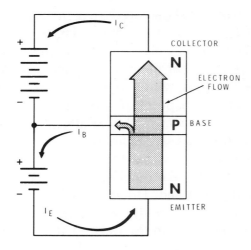

Figure 3-15
Electron flow in an NPN transistor.

Figure 3-15 shows the electron flow in a correctly biased NPN transistor. A large number of electrons enter the emitter N-region due to the forward biased emitter-base junction. Since the base region is designed to be very thin, the high positive potential on the collector causes most electrons to pass straight through the base. Of course, some will remain in the base, giving a small base current.

An important relationship exists between emitter current (I_E), base current (I_B), and collector current (I_C). It can be stated mathematically as follows:

$$I_E = I_B + I_C$$

The equation simply states that the emitter current (I_E) is equal to the sum of the base current (I_B) and the collector current (I_C). Another way of looking at this is that the collector current is equal to the emitter current less the current lost to the base.

$$I_C = I_E - I_B$$

If the base region of the transistor was not extremely thin, the action just described could not occur. The thin base region makes it possible for the emitter injected electrons to move quickly into the collector region. A large base region would minimize the interaction between the emitter and collector regions and the transistor would act more like two separately biased diodes.

It is interesting to note that a small change in I_B causes a large change in I_C. For example if I_B is decreased, fewer electrons cross the emitter-base junction. Therefore, fewer electrons cross the collector-base junction, decreasing I_C. On the other hand if I_B is increased, more electrons cross the emitter-base junction, which in turn increases I_C. Thus, I_B controls I_C. To further dramatize this point, if I_B is reduced to zero, I_C also drops to zero because there is no electron flow from emitter to base. This condition is known as **cutoff**, since there is no transistor current flow. Likewise, I_B can be increased to a point where no more electrons are available from the emitter and, therefore, I_C cannot increase. This condition is called **saturation** and is usually avoided. However, when a transistor is used as a switch, such as in digital circuits, it generally alternates between cutoff and saturation depending on the desired state of the switch.

While the preceding discussion dealt with the NPN transistor, the PNP is very similar. However, the current within a PNP transistor is in terms of "hole" flow, as shown in Figure 3-16. This is due to the P-type emitter and collector whose current or charge "carriers" are holes. Note also that, for correct bias, the batteries must be reversed. This means that the external current will be in the opposite direction to an NPN transistor.

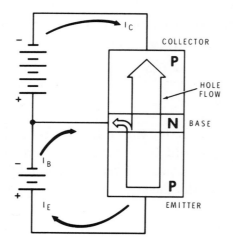

Figure 3-16
Hole flow in a PNP transistor.

Transistor Amplification

Now that you have seen basic transistor operation, the next step is transistor amplification. We will continue the discussion of the NPN transistor of Figure 3-15, however, we will use the transistor's schematic symbol as shown in Figure 3-17. Notice that the voltage sources used to forward bias the emitter junction and reverse bias the collector junction are labeled V_{EE} and V_{CC} respectively. These two symbols are standard designations for the voltages used in transistor circuits. Furthermore, V_{EE} is variable so that the forward bias voltage can be varied. The transistor is designated Q, since the Q designation is widely used to represent transistors.

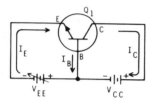

Figure 3-17

Schematic diagram of a properly biased NPN transistor.

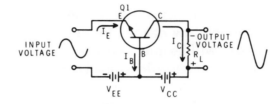

Figure 3-18

A basic NPN transistor amplifier circuit.

In order for the transistor shown in Figure 3-17 to provide amplification, the device must be capable of accepting an input signal (current or voltage) and providing an output signal that is greater in strength or amplitude. The transistor cannot perform this function if it remains connected as shown in Figure 3-17. Instead, the circuit must be changed so that it appears as shown in Figure 3-18. Notice the resistor between the collector of transistor Q_1 and the positive terminal of V_{CC}. This component is commonly referred to as a load resistor and is labeled R_L. It is used to develop an output voltage with the polarity indicated. In other words, the collector current must flow through this resistor to produce a specific voltage drop. Note also the separation between the emitter of transistor Q_1 and the negative terminal of V_{EE}. These open connections serve as the input of the amplifier circuit and allow an input signal to be inserted between the emitter of Q_1 and V_{EE}. Therefore, the input is not an open circuit, but is completed when an external voltage source is connected between the emitter and V_{EE}. Furthermore, the values of V_{EE} and V_{CC} have been selected to bias transistor Q_1 so that the values of I_E, I_B and I_C are high enough to permit proper circuit operation.

The emitter junction of transistor Q_1 is forward-biased and like a PN junction diode, has a relatively low resistance. However, the collector junction is reverse-biased and therefore has a relatively high resistance. In spite of the tremendous difference in emitter junction resistance and collector junction resistance the emitter current (I_E) is almost equal to the collector current (I_C). The value of I_C is only slightly less than I_E because a small portion of I_E flows out of the base region to become base current (I_B). The load resistor can therefore have a high value without greatly restricting the value of I_C.

If a small voltage is applied to the input terminals which aids the forward bias voltage (V_{EE}), the value of V_{EE} is effectively increased. This will cause I_E to increase and I_C will also increase. This will increase the voltage drop across the load resistor (R_L). However, when the input voltage is reversed so that it opposes V_{EE}, the value of V_{EE} is effectively reduced and both I_E and I_C will decrease. This will cause the voltage drop across R_L to decrease. A change in input voltage can, therefore, produce a corresponding change in output voltage. However, the output voltage change will be much greater than the input change. This is because the output voltage is developed across a high output load resistance (R_L) and the input voltage is applied to the low resistance offered by the emitter junction. Thus, a very low input voltage can control the value of I_E which in turn determines the value of I_C. Even though I_C remains slightly less than I_E, it is forced to flow through a higher resistance and, therefore, produce a higher output voltage.

The action just described will occur if the input signal is either a DC or an AC voltage. The important point to remember is that any change in input voltage is greatly amplified by the circuit so that a much larger but proportional change in output voltage is obtained. It is also important to realize that the input signal is not simply increased in size, but is used to control the conduction of transistor Q_1 and this transistor in turn controls the current through the load resistor (R_L). The transistor is therefore made to take energy from the external power source V_{CC} and apply it to the load resistor in the form of an output voltage whose value is controlled by a small input voltage.

The transistor shown in Figure 3-18 is therefore being used to convert a relatively low voltage to a much higher voltage. Any circuit which performs this basic function is commonly referred to as a **voltage amplifier**. Such circuits are widely used in electronic equipment since it is often necessary to raise both DC and AC voltages to higher levels so that they may be effectively used. Later in this unit you will also discover that the transistor can be connected in a different manner which will allow it to amplify currents as well as voltages. Any circuit which is used for the sole purpose of converting a low current to a higher current is commonly referred to as a **current amplifier**.

Although an NPN transistor was used in the amplifier circuit shown in Figure 3-18, the PNP transistor will perform the same basic function. However; in the PNP circuit the bias voltages (V_{EE} and V_{CC}) must be reversed and the respective currents (I_E, I_B, and I_C) will flow in the opposite directions.

Transistor Circuit Arrangements

As explained previously, the transistor is used primarily as an amplifying device. However, there is more than one way of using a transistor to provide amplification and each method offers certain advantages or benefits but also has certain disadvantages or limitations. Basically, the transistor may be utilized in three different circuit arrangements to perform its important amplifying function. In each arrangement one of the transistor's three leads is used as a common reference point and the other two leads serve as input and output connections. The three circuit arrangements are referred to as the **common-base** circuit, the **common-emitter** circuit and the **common-collector** circuit, and each arrangement may be constructed by using either NPN or PNP transistors. In many cases the common leads are connected to circuit or chassis ground and the circuit arrangements are then often referred to as grounded-base, grounded emitter, and grounded-collector circuits. Furthermore, in each circuit arrangement the transistor's emitter junction is always forward-biased while the collector junction is reverse-biased.

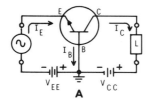

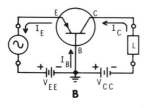

Figure 3-19
Common-base circuits.

COMMON-BASE CIRCUITS

In the common-base circuit, the transistor's base region is used as a common reference point and the emitter and collector regions serve as the input and output connections. This basic circuit arrangement is shown in Figure 3-19. Figure 3-19A shows how an NPN transistor is connected in the common-base configuration while Figure 3-19B shows how the same arrangement is formed with a PNP transistor. Notice that both circuits are arranged in the same basic manner, but the polarity of the bias voltages (V_{EE} and V_{CC}) are opposite so that the NPN and PNP transistors are properly biased. This common-base arrangement was also described in the previous section. This arrangement was used because it serves as an excellent model for explaining basic transistor operation.

In the common-base circuit the input signal is applied between the transistor's emitter and base and the output signal appears between the transistor's collector and base. The base is therefore common to both the input and output. This input signal applied to the circuit is represented by the AC generator symbol. The output voltage is usually developed across a resistive component but in some cases components which have inductive as well as resistive properties (coils, relays, or motors) may be used. In general, the component that is connected to the output of the circuit is referred to as a *load*. In Figure 3-19 the load is identified by the box which contains the letter "L".

The common-base circuit is useful because (as explained earlier) it provides voltage amplification. However, this circuit has other basic characteristics which should be considered. Since the emitter junction of the transistor is forward-biased the input signal sees a very low emitter-to-base resistance or in other words the transistor has a low input resistance. However, the collector junction of the transistor is reverse-biased and therefore offers a relatively high resistance. The resistance across the output terminals (collector-to-base) of the transistor is therefore much higher than the input resistance. The input resistance of a typical low power transistor that is connected in the common-base configuration could be as low as 30 or 40 ohms; however, the output resistance of the transistor could be as high as 1 megohm.

The common-base circuit (as explained previously) provides slightly less output current than input current because I_C is slightly lower than I_E. In other words there is a slight current loss between the input and output terminals of the circuit. However, the circuit is still useful as a voltage amplifier as explained earlier. Since the input and output currents are almost the same and because the output voltage developed across the load can be much higher than the input signal voltage, the output power produced by the circuit is much higher than the input power applied to the circuit. The common-base circuit therefore provides power amplification as well as voltage amplification.

COMMON-EMITTER CIRCUITS

Transistors (NPN and PNP) may also be connected as shown in Figure 3-20. Notice that the input signal is applied between the base and emitter of each transistor while the output signal is developed between the collector and emitter of each transistor. Since the emitter of each transistor is common to the input and output of each circuit, the arrangements are referred to as common-emitter circuits. Notice that the emitter junction of each transistor is still forward-biased while the collector junction of each device is reverse-biased. Also, the NPN circuit shown in Figure 3-20A and the PNP circuit shown in Figure 3-20B require bias voltages that are opposite in polarity. Note that the forward bias voltage (designated as V_{BB}) is now applied to the base of the transistor instead of the emitter. In other words the forward bias voltage (V_{BB}) is applied so that it controls the base current (I_B) instead of the emitter current (I_E). Only the base current flows through bias voltage V_{BB} and the AC generator (signal source). The reverse bias voltage (V_{CC}) is applied through the output load to the collector and emitter regions of the transistor and only the collector current (I_C) flows through the load and V_{CC}.

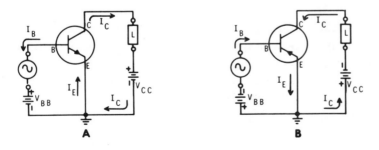

Figure 3-20
Common-emitter circuits.

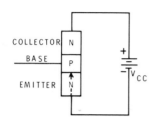

Figure 3-21
The reverse-biased collector
junction in an NPN
common-emitter circuit.

Although it may not be apparent, V_{CC} does actually reverse bias the collector-base junction of the transistor. Figure 3-21 shows how this reverse bias is obtained in an NPN common-emitter circuit. If V_{CC} is connected across the emitter and collector regions of the transistor as shown, the emitter-base junction becomes forward-biased and therefore has a low resistance. However, V_{CC} causes the N-type collector to be positive with respect to the P-type base and the collector-base junction is therefore reverse-biased. The forward-biased emitter-base junction effectively allows the negative terminal of V_{CC} to be applied to the P-type base region. V_{CC} is therefore effectively applied to the collector and base regions as shown. If V_{CC} was reversed, the collector junction would become forward-biased and the emitter junction would become reverse-biased; however, this condition would be undesirable. The same basic action also occurs in a PNP transistor, although the PNP device requires bias voltage polarities which are exactly opposite to those used with an NPN device.

The three currents (I_E, I_B, and I_C) in the common-emitter circuit still have the same relationship as before and still vary in a proportional manner. In other words I_E is still equal to $I_B + I_C$ even though I_B has now become the input current and I_C has become the output current. Any change in the input current (I_B) will therefore result in a proportional change in I_E and I_C. The difference between this circuit and the common-base circuit previously described is therefore simply one of reference. We have now connected the transistor so that its very low base current is being used to control its much larger collector current.

When an input signal voltage is applied to the common-emitter circuit, it either aids or opposes V_{BB} and therefore causes I_B to either increase or decrease. The much higher emitter and collector currents (which are almost equal) are forced to increase and decrease by the same percentage. Since a very low I_B is used to control a relatively high I_C, the transistor is now being used to provide current amplification as well as voltage amplification. This of course means that the power supplied to the load is much greater than the power applied to the input of the circuit. In fact the common-emitter arrangement provides the highest power output for a given power input when compared with the other two circuit arrangements. The common-emitter circuit is also the only circuit that provides current, voltage, and power amplification.

A transistor that is used in a common-emitter arrangement will also have a low input resistance but this resistance will not be as low as the input resistance of a common-base arrangement. This is because the input base current for the common-emitter circuit is much lower than the input emitter current of the common-base circuit. Also, the output resistance of a transistor connected in a common-emitter arrangement will be somewhat lower than when it is connected in a common-base circuit. A typical low power transistor might have an input resistance of 1000 or 2000 ohms and an output resistance of 50,000 or 60,000 ohms.

COMMON-COLLECTOR CIRCUITS

The NPN and PNP transistors shown in Figure 3-22 are connected so that the collector of each transistor is used as a common reference point instead of the emitter and the base and the emitter regions serve as the input and output connections. These circuits are known as common-collector circuits.

In the common-collector circuit the input signal voltage is applied between the base and collector regions of the transistor. The input either aids or opposes the transistor's forward bias and therefore causes I_B to vary accordingly. This in turn causes I_E and I_C to vary by the same percentage. An output voltage is developed across the load which is connected between the emitter and collector regions of the transistor. The emitter-current (I_E) flowing through the load is much greater than the base current (I_B) and the circuit therefore provides an increase in current between its input and output terminals. However, the voltage developed across the load will always be slightly lower than the voltage applied to the circuit. The slightly lower voltage will appear at the emitter of the transistor because the device tends to maintain a relatively constant voltage drop across its emitter-base junction. As explained earlier, this forward voltage drop may be equal to approximately 0.3 volts if the transistor is made of germanium or 0.7 volts if the transistor is made of silicon. The output voltage appearing at the emitter of the transistor therefore tends to track or follow the input voltage applied to the transistor's base. For this reason the common-collector circuit is often called an **emitter follower**.

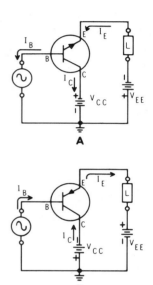

Figure 3-22
Common-collector circuits.

The common-collector circuit therefore functions as a current amplifier but does not produce an increase in voltage. However, the increase in output current (even though output voltage is slightly less than input voltage) results in a moderate increase in power. Also, the input resistance of any transistor connected in a common-collector arrangement is extremely high. This is because the input resistance is that resistance which appears across the reverse-biased collector-base junction. This input resistance can be as high as several hundred thousand ohms in a typical low power transistor. However, the output resistance appearing between the transistor's emitter and collector regions will be much lower (often as low as several hundred ohms) because of the relatively high emitter current (I_E) that flows through the output lead.

The common-collector circuit is not useful as an amplifying device. Instead, this circuit is widely used in applications where its high input resistance and low output resistance can perform a useful function. The circuit is often used to couple high impedance sources to low impedance loads and therefore perform the same basic function as an impedance matching transformer. The circuit also provides a moderate amount of power gain.

The bias voltages (V_{BB}, V_{EE}, and V_{CC}) used in the three circuit arrangements just described may appear to restrict or interfere with the flow of signal current through these circuits; however, this does not occur. The bias voltages are effectively shorted as far as the input and output signals are concerned, although in some applications large capacitors are placed across the voltage sources to insure that they offer minimum impedance to signal currents. For example, a large electrolytic capacitor could be placed across V_{CC} in Figure 3-22 to insure that the collector of the transistor is at ground potential as far as AC signals are concerned.

SUMMARY

As can be seen from the previous discussion, each transistor circuit configuration has specific characteristics. The chart shown in Figure 3-23 compares the characteristics of these three circuits. As you can tell, the common-emitter configuration is the most versatile: offering voltage, current, and power gain. It is also the most widely used configuration. However, the common-base and common-collector circuits have specific applications, such as voltage amplification and impedance matching, respectively.

CHARACTERISTICS	TYPICAL VALUES		
	COMMON — EMITTER	COMMON — BASE	COMMON — COLLECTOR
IMPUT IMPEDANCE	1000-2000 Ω	30-40 Ω	Up to 500 kΩ
OUTPUT IMPEDANCE	50-60 kΩ	Up to 1 MΩ	400 Ω
VOLTAGE GAIN	Up to 500	Up to 1,000	Less than 1
CURRENT GAIN	30	Less than 1	50
POWER GAIN	Up to 10,000	Up to 1,000	50

Figure 3-23
Transistor amplifier configuration characteristics.

Transistor Ratings

One of the more important transistor ratings is the forward current transfer ratio or beta (β). **Beta** is the transistor's current gain in the common-emitter configuration. It is the ratio of a small change in collector current to a small change in base current. When expressed as a formula, we have:

$$\beta = \frac{\Delta I_C}{\Delta I_B}$$

where
β = beta
ΔI_C = A small change in collector current
ΔI_B = A small change in base current

Typical values of β range from 20 to 200.

One disadvantage of transistors when compared to vacuum tubes is their sensitivity to temperature. When the semiconductor material is heated, more electrons are freed due to thermal agitation. This allows more current flow in the transistor. The additional current creates additional heat! If this cycle continues, the transistor will be in **thermal runaway** and will soon be destroyed. External bias circuitry must be designed to compensate for any heat changes to prevent thermal runaway.

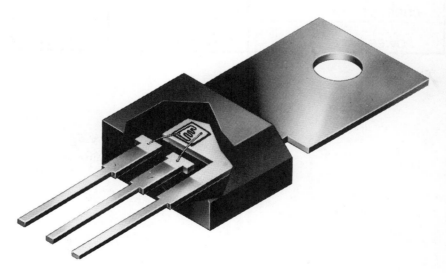

Figure 3-24
A cutaway view of a high power transistor.

A rating which is related to a transistor's thermal characteristic is its **maximum power dissipation**. This rating is determined by how much heat a transistor can safely dissipate. The input power to a transistor is determined by multiplying collector current by collector voltage. Its power dissipation (P_{dis}) is then input power (P_{in}) minus output power (P_{out}) or

$$P_{dis} = P_{in} - P_{out}$$

Therefore, any input power which is not transferred to the output or load is dissipated as heat inside the transistor.

One way to increase a transistor's maximum power dissipation is to construct it so that it can radiate heat more effectively. A transistor capable of dissipating up to 30 watts is shown in Figure 3-24. It has a special tab which can be attached to an external heat sink. The tab "conducts" heat away from the transistor to the heat sink. The heat sink has many fins which radiate the heat into the air. Figure 3-25 shows the heat sink for a transistorized 100-watt amateur radio transceiver.

HEAT SINK

Figure 3-25
The heat sink on a 100 watt solid state
amateur radio transceiver.

Self-Review Questions

8. Identify the type of transistor and the transistor leads by filling in the blanks in Figure 3-26.

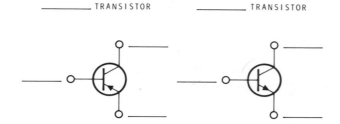

Figure 3-26
Fill in the blanks.

9. In a correctly biased transistor, the emitter-base junction is _____ biased, while the collector-base junction is
    <u>forward/reverse</u>

    _____ biased.
    <u>forward/reverse</u>

10. Name the amplifier configurations shown in Figure 3-27.

    A. _____

    B. _____

    C. _____

11. The common-collector circuit has _____ input impedance,
    <u>low/high</u>

    _____ output impedance, and high _____ gain.
    <u>low/high</u>

12. What is the primary application for the common-collector configuration? _____

    _____

13. The common-base circuit has _____ input impedance,
    <u>low/high</u>

    _____ output impedance, and high _____ gain.
    <u>low/high</u>

14. The common-_____ circuit provides the highest power amplification.

15. Regardless of the type of circuit arrangement used, the transistor's emitter junction must always be _____-biased.

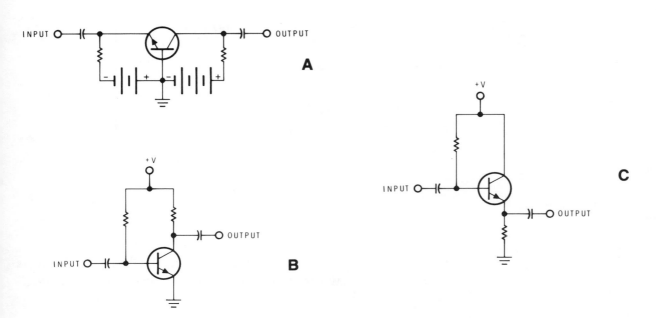

Figure 3-27
Identify these amplifier configurations.

16. The ratio of a small change in base current to a small change in collector current is called _____ .

17. What is thermal runaway? _____
_____
_____

18. A transistor's construction and its ability to effectively radiate heat determine its maximum _____ _____ .

Self-Review Answers

8.   See Figure 3-28.

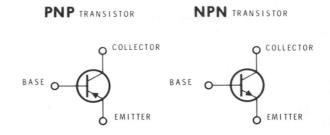

Figure 3-28

9.   In a correctly biased transistor, the emitter-base junction is **forward**-biased, while the collector-base junction is **reverse**-biased.

10.   Even though you have not seen the circuits of Figure 3-27 before, you should be able to identify the configuration from the input and output connections. They are as follows:

   A.   common-base

   B.   common-emitter

   C.   common-collector.

11.   The common collector circuit has **high** input impedance, **low** output impedance, and high **current** gain.

12.   **Impedance matching** is the primary application for the common collector amplifier.

13. The common-base circuit has **low** input impedance, **high** output impedance, and high **voltage** gain.

14. The common-**emitter** circuit provides the highest power amplification.

15. Regardless of the type of circuit arrangement used, the transistor's emitter junction must always be **forward**-biased.

16. The ratio of a small change in base current to a small change in collector current is called beta or β.

17. Thermal runaway occurs when a transistor heats up, which causes an increase in current. The increased current causes additional heat. Eventually the transistor will be destroyed. Thermal runaway is prevented by using a bias network that compensates for heat changes.

18. A transistor's construction and its ability to effectively radiate heat determine its maximum **power dissipation**.

FIELD-EFFECT TRANSISTORS

The field-effect transistor (FET) is a three terminal solid-state device capable of amplification when used in the proper circuitry. Its three terminals are called the **source**, **drain**, and **gate**. These terminals roughly correspond to the emitter, collector, and base, respectively, of the junction transistor. However, in the FET the source to drain current is controlled by a **voltage** applied to the gate. Compare this to the junction transistor which is a current controlled device. Also, since no current flows into the gate, the input impedance is very high, ranging from 10^6 to 10^{14} Ω.

There are two basic types of field-effect transistors: the junction FET or JFET, and the insulated gate FET or MOSFET. You'll be studying both types in this section.

Junction FET

The basic construction of a junction FET is shown in Figure 3-29A. It consists of a block of N-type semiconductor to which the source and drain leads are attached as shown. This N-type material is called the **channel**, and the electrons flow through this material from source to drain. The gate is a slice of P-type semiconductor that forms a junction with the N-type channel.

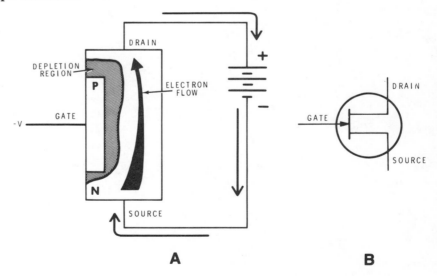

A   **B**

Figure 3-29
N-channel Junction FET, construction (A),
schematic symbol (B).

Under normal conditions the gate junction is reverse-biased. This is accomplished by applying a negative bias voltage to the gate. The resulting depletion region will extend across the channel and, since the depletion region is "depleted" of electrons, it will limit the current flow through the channel. Therefore, the channel current is controlled by varying the amount of reverse bias on the gate junction. In fact, the gate voltage can be increased until the depletion region covers the entire N-channel, thus, cutting-off all current flow. This condition is called **pinch-off**. Since the gate junction is always reverse-biased, the JFET's input impedance is very high.

The schematic symbol of the JFET is shown in Figure 3-29B. Note that the gate lead has an arrow. It points to the N-type material, just as in the junction transistor and diode. Figure 3-29 shows an N-channel JFET while Figure 3-30 shows a P-channel JFET. The only differences are the positive bias voltage required to reverse-bias the gate junction, and the positive charge carriers or holes in the channel. The operation is the same as the N-channel JFET. Note the schematic symbol in Figure 3-29B.

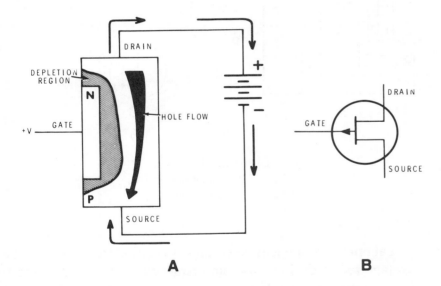

A   **B**

Figure 3-30
P-channel Junction FET, construction (A),
schematic symbol (B).

Insulated Gate FET

The construction of an insulated gate FET is shown in Figure 3-31A. In this type of FET the channel is an N-type semiconductor that is placed on a P-type **substrate**. The substrate is a block of semiconductor material upon which the rest of the FET is constructed. The gate is a strip of metal which is insulated from the channel by a layer of metal oxide. For this reason the insulated gate FET is also known as a metal oxide semiconductor FET or MOSFET. Since the gate is insulated from the channel the input impedance is extremely high, typically 10^{10} to 10^{14} Ω.

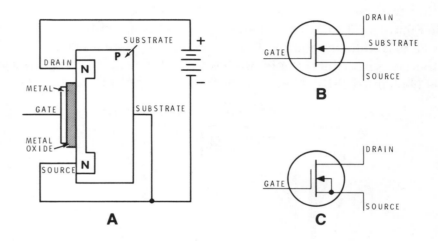

Figure 3-31
Insulated gate FET, construction (A),
schematic symbols (B and C).

During normal operation a negative voltage is applied to the source and a positive voltage to the drain. This establishes current flow through the channel from source to drain. If a negative voltage is applied to the gate, the negative charge on the gate will repel the electrons in the channel and a depletion region will form. The gate potential will control the size of the depletion region and, therefore, the channel current flow. This is very similar to the junction FET. However, when a positive voltage is applied to the MOSFET gate, current flow in the channel is enhanced. That is, electrons are drawn toward the gate from the channel and the substrate. This increases channel conductivity and, thus current flow. This could never be done in a JFET because the gate junction would become forward biased, allowing gate current and drastically reducing input impedance. However, the MOSFET gate is insulated and its input impedance remains high regardless of gate voltage polarity. Thus, the MOSFET's source to drain current is controlled by a positive or negative voltage applied to the gate.

The insulated gate FET schematic symbol is shown in Figure 3-31B. Note that the gate is shown insulated from the channel. The substrate lead is identified by the arrow and, as always, the arrow points toward the N-type semiconductor. In this case the arrow identifies the device as an N-channel MOSFET. P-channel devices are also available and their operation is similar. In the device shown, the substrate has a separate lead, however, in many MOSFETs the substrate is connected to the source internally. This is shown in Figure 3-31C.

Another type of MOSFET is shown in Figure 3-32A. Note that the construction is similar to that just discussed except that it does not have a complete channel. The source and drain connections are there but a complete path for electron flow does not exist.

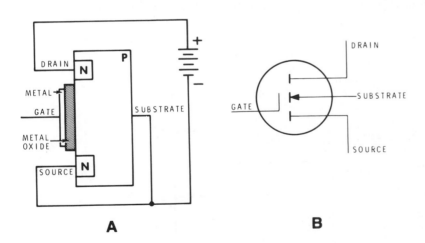

Figure 3-32
Enhancement mode MOSFET, construction (A),
schematic symbol (B).

Therefore, with no bias voltage applied to the gate, source to drain current is zero. However, when a positive voltage is applied to the gate, the substrate electrons are attracted toward it. This actually induces a conduction path or channel from the source to the drain. The amount of positive voltage on the gate determines the channel width and, therefore, the source to drain current. This type of device is known as an **enhancement mode** MOSFET since electron flow is "enhanced" by the application of a positive gate voltage. The previously discussed device is called a **depletion mode** MOSFET since electron flow is "depleted" by a negative gate voltage.

The schematic symbol for an enhancement mode MOSFET is shown in Figure 3-32B. Note the broken line indicating the type of channel and the substrate arrow showing that it is an N-channel device. P-channel units are also available.

One problem encountered with insulated gate FETs is that too high a gate voltage may actually puncture the insulating layer, thus destroying the FET. Because of the extremely high gate resistance, even static voltages from your fingertips can puncture the oxide layer. Therefore, to avoid damage, MOSFETs are shipped with their leads shorted together. Another widely used method of protection is to wire back-to-back zener diodes across the gate connection. This is done internally as shown in Figure 3-33. The zener diodes place an upper limit on the gate voltage since they will conduct when their zener voltage is reached. Thus, the voltage applied to the gate can never exceed the zener voltage.

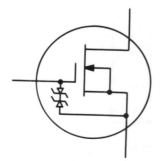

Figure 3-33
A static-voltage protected depletion mode MOSFET.

FET Circuit Arrangements

Like bipolar transistors, FET's are used primarily to obtain amplification and like bipolar transistors, FET's can be connected in three different circuit arrangements. These three configurations are commonly referred to as **common-source, common-gate,** and **common-drain** circuits. The circuits' connections are valid for both JFET's and IGFET's.

COMMON-SOURCE CIRCUITS

The common-source circuit is the most widely used FET circuit arrangement. This circuit configuration is comparable to the common-emitter circuit arrangement that was described in a previous section. A basic common-source configuration is shown in Figure 3-34. Notice that the input signal is applied between the gate and source leads of the FET and the output signal appears between the drain and source leads. The source is therefore common to both input and output.

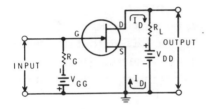

Figure 3-34
Basic common-source circuit.

An N-channel junction FET is shown in Figure 3-34 and it is therefore biased so that its gate is negative with respect to its source. The gate-to-source bias voltage is provided by an external voltage source (designated as V_{GG}) which is in series with a resistor (R_G). Therefore, V_{GG} is not applied directly to the gate and source leads. However, the FET's gate-to-channel PN junction is reverse-biased so that no current can flow through the gate lead and R_G. The voltage across R_G is therefore zero and the full value of V_{GG} is effectively placed across the gate and source leads. Normally R_G has a high value of resistance (often more than 1 megohm) so that the resistance seen at the input of the circuit will remain high. A low value of resistance could reduce the input resistance of the circuit since R_G is effectively in parallel with the FET's high input (gate-to-source) resistance as far as the input signal is concerned.

The external bias voltage (V_{GG}) is adjusted so that a specific value of drain current (I_D) will flow through the FET and the device will operate correctly. The value of I_D is also controlled (but to a lesser extent) by external voltage source V_{DD}. This voltage source supplies the necessary drain-to-source operating voltage for the FET but another resistor (R_L) is inserted between V_{DD} and the FET's drain lead. The drain current flowing through the FET therefore flows through R_L thus causing a voltage to appear across this resistor. A portion of the voltage (V_{DD}) is therefore dropped across the FET and the rest appears across R_L.

Since the FET is controlled by an input voltage (not an input current) the common-source arrangement is used to obtain voltage amplification. For example, an AC input voltage will alternately aid and oppose the input bias voltage (V_{GG}) so that the FET's gate-to-source voltage will vary with the changes in input voltage. This will in turn cause the FET to alternately conduct more and less drain current. The FET therefore acts like a variable resistor in series with a fixed resistor (R_L) and the resistance of the FET is effectively controlled by the input signal voltage. As the FET conducts more and less drain current, its drain-to-source voltage varies accordingly to produce an output voltage that changes in response to the input voltage. However; by making R_L relatively large (often more than 10 k ohms) and by biasing the FET so that its drain-to-source resistance is also high, the changes in drain current (even when small) can produce an output signal voltage that is much higher than the input signal voltage.

In addition to providing voltage amplification, the common-source circuit has another desirable feature. Due to the extremely high gate-to-source resistance of the FET, the circuit has a very high input resistance even though input resistor R_G is used. This means that the common-source circuit will usually have a minimum loading effect on its input signal source. Common-source circuits are therefore widely used in digital or computer applications where a number of circuit inputs must often be connected to the output of one circuit without affecting its operation.

The output resistance of the common-source circuit is lower than its input resistance because of the lower source-to-drain resistance of the FET. However, the circuit still has a moderately high output resistance, since the FET is usually biased to conduct a relatively low drain current which is often not more than a few milliamperes.

The common-source circuit may also be used to amplify both low frequency and high frequency AC signal voltages as well as a wide range of DC signal voltages. For example, this circuit is often used in the input rf amplifier stage of radio receivers to amplify the wide range of high frequency input signals which can vary widely in amplitude. The circuit is also used in electronic test instruments such as solid-state voltmeters or multimeters. In these applications it must amplify a wide range of DC or AC voltages and at the same time present a high input resistance to prevent undesirable loading of the circuits being tested.

Although the common-source circuit shown in Figure 3-34 is formed with an N-channel junction FET, the same basic circuit can be formed with a P-channel JFET if the polarities of the bias voltages are reversed. Also, this basic circuit can be formed with depletion-mode and enhancement-mode MOSFET's. When these insulated-gate devices are used, the additional substrate leads are generally connected to their source leads or to circuit ground. Also, it is important to note that depletion-mode MOSFET's can accept positive and negative gate voltages and may be operated with zero gate bias, while enhancement-mode devices require a gate bias voltage in order to conduct. This means that slightly different biasing arrangements must be used to accommodate the various FET types when the basic common-source arrangement is used.

COMMON-GATE CIRCUITS

The common-gate circuit may be compared to the common-base transistor circuit because the electrode which has primary control over the FET's conduction (its gate) is common to the circuit's input and output. A basic common-gate circuit is shown in Figure 3-35. An N-channel JFET is used in this circuit and it therefore requires the same basic operating voltages as the N-channel device in the common-source circuit previously described. However, in this circuit the input gate-to-source bias voltage is provided by voltage V_{SS} and resistor R_S while the output portion of the circuit is biased by the voltage V_{DD} and resistor R_L.

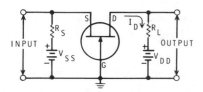

Figure 3-35
Basic common-gate circuit.

An AC input voltage will effectively vary the FET's gate-to-source voltage and cause variations in its conduction. The FET's drain current flows through R_L and variations in this current produces voltage changes across the FET which follow the input signal. The voltage developed across the drain and gate leads of the FET serves as the output.

Like the common-source circuit, the common-gate circuit provides voltage amplification; however the voltage gain of the common-gate arrangement is lower. The input resistance of the circuit is also low since current flows through the input source lead; however the output resistance of the circuit is relatively high. This makes the common-gate circuit suitable in applications where a low resistance generator must supply power to a high resistance load. When inserted between the generator and load, the common-gate circuit effectively matches the low and high resistances to insure an efficient transfer of power.

Although it has a low voltage gain, the common-gate circuit is often used to amplify high frequency AC signals. This is because of the circuit's low input resistance and because the circuit inherently prevents any portion of its output signal from feeding back and interfering with its input signal. The circuit is therefore inherently stable at high frequencies and additional components are not required to prevent interference between input and output signals. Such stabilizing components are often required with the common-source circuit arrangement previously described to insure reliable operation.

The common-gate circuit may also be formed with a P-channel JFET or with depletion-mode or enhancement-mode MOSFET's. However, when an MOSFET is used, the substrate lead is usually connected directly to the gate or to circuit ground and additional components may be required to properly bias the device.

COMMON-DRAIN CIRCUITS

The common-drain circuit is similar to the common-collector bipolar transistor circuit. This basic FET circuit arrangement is shown in Figure 3-36. The N-channel JFET in this circuit receives its gate-to-source bias voltage from V_{GG} and R_G. The output (drain-to-source) portion of the FET is biased by V_{DD} and R_L. The input signal voltage is effectively applied between the FET's gate and drain even though it appears to be applied to the gate and source leads. The output signal is effectively developed across the FET's source and drain even though it appears to be taken directly from R_L. This is because voltage source V_{DD} is effectively a short as far as the input and output signals are concerned. Therefore, V_{DD} effectively grounds the FET's drain and makes it common to both input and output signals.

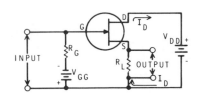

Figure 3-36
Basic common-drain or source follower circuit.

The common-drain circuit cannot provide voltage amplification. The voltage appearing at the source of the FET tends to track or follow the voltage at the gate and for this reason the circuit is often referred to as a **source-follower**. However the output source voltage is always slightly less than the input gate voltage thus causing the circuit to have a voltage gain that is less than one or unity.

The input resistance of the common-drain circuit is extremely high. In fact, this circuit has a higher input resistance than a common-source or a common-gate configuration. However, the output resistance of this circuit is low. This makes the common-drain circuit suitable for coupling a high resistance generator to a low resistance load so that an efficient transfer of power can take place.

The common-drain configuration can also be formed with P-channel JFET's as well as depletion-mode and enhancement-mode IGFET's.

Self-Review Questions

19. The junction transistor is a _____ controlled device, while the FET is a _____ controlled device.

20. In an FET electrons flow from source to drain through the _____ .

21. The major advantage of the FET is its very high _____ _____ .

22. Identify the type of FET and the terminals by filling in the blanks in Figure 3-37.

23. With no gate bias the depletion mode MOSFET is always _____ while the enhancement mode MOSFET is always
   on/off
   _____ .
   on/off

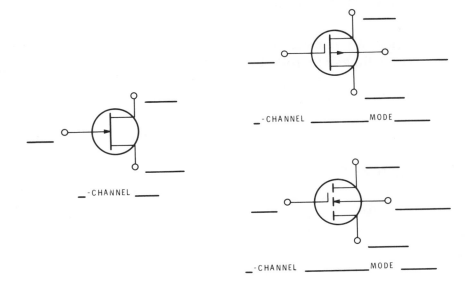

Figure 3-37
Fill in the blanks.

24. How can static voltages damage MOSFETS? _____
   _____
   _____
   _____

25. The most widely used FET circuit arrangement is the
   _____-_____ circuit.

26. The FET circuit arrangement that has the highest voltage gain is the
   _____-_____ circuit.

27. The FET circuit arrangement that inherently provides the most
   stable operation at high signal frequencies is the _____-
   _____ circuit.

28. What FET circuit arrangement can be used to match a high impedance generator to a low impedance load? _____
   _____

29. What FET circuit arrangement can be used to match a low impedance generator to a high impedance load? _____

Self-Review Answers

19.   The junction **transistor** is a **current** controlled device, while the FET is a **voltage** controlled device.

20.   In an FET, electrons flow from source to drain through the **channel**.

21.   The main advantage of the FET is its very high **input impedance**.

22.   See Figure 3-38.

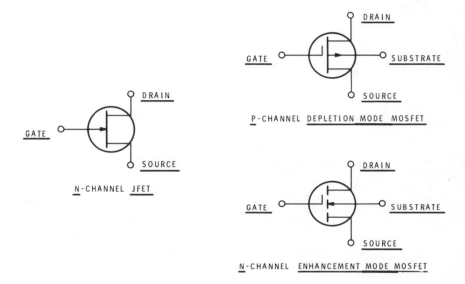

Figure 3-38

23. With no gate bias the depletion mode MOSFET is always **on** while the enhancement mode is always **off**.

24. Static voltages can puncture the oxide layer of MOSFETs and destroy the device.

25. The most widely used FET circuit arrangement is the **common-source** circuit.

26. The FET circuit arrangement that has the highest voltage gain is the **common-source** circuit.

27. The FET circuit arrangement that inherently provides the most stable operation at high signal frequencies is the **common-gate** circuit.

28. A **common-drain** or **source-follower** circuit can be used to match a high impedance generator to a low impedance load.

29. A **common-gate** circuit can be used to match a low impedance generator to a high impedance load.

OPTOELECTRONIC DEVICES

In this section you will examine a group of solid-state components which are capable of converting light energy into electrical energy or electrical energy into light energy. These components are commonly referred to as **optoelectronic** devices since their operation relies on both optic and electronic principles.

The optoelectronic devices discussed in this section are divided into two basic groups. They are classified as light-sensitive devices and light-emitting devices. The most important components found within these two categories are described in detail and various applications are considered.

In order to understand the operation of optoelectronic devices, it is first necessary to understand the basic principles of light. Therefore, we will begin this section by defining light and considering its various properties.

Basic Characteristics of Light

The term *light* is used to identify *electromagnetic radiation* which is visible to the human eye. Basically, light is just one type of electromagnetic radiation and differs from other types such as cosmic rays, gamma rays, X-rays, and radio waves only because of its frequency.

The light spectrum extends from approximately 300 gigahertz* to 300,000,000 gigahertz. It is wedged midway between the high end of the radio frequency (RF) waves, which roughly extend up to 300 gigahertz, and the X-rays, which begin at roughly 300,000,000 gigahertz. Above the X-ray region are the gamma rays and then the cosmic rays.

*One gigahertz, abbreviated GHz, is equal to a frequency of 1,000,000,000 Hertz or 10^9 cycles per second.

Within the 300 to 300,000,000 GHz light spectrum only a narrow band of frequencies can actually be detected by the human eye. This narrow band of frequencies appears as various colors such as red, orange, yellow, green, blue and violet. Each color corresponds to a very narrow range of frequencies within the visible region. The entire visible region extends from slightly more than 400,000 GHz to approximately 750,000 GHz. Above the visible region (between 750,000 and 300,000,000 GHz) the light waves cannot be seen. The light waves which fall within this region are referred to as **ultraviolet rays**. Below the visible region (between 300 and 400,000 GHz) the light waves again cannot be seen. The light waves within this region are commonly referred to as **infrared rays**. The entire light spectrum is shown in Figure 3-39.

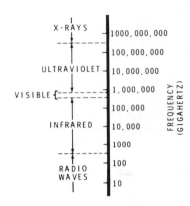

Figure 3-39
The light spectrum.

Although light is assumed to propagate or travel as electromagnetic waves, wave theory alone cannot completely explain all of the phenomena associated with light. For example, wave theory may be used to explain why light bends when it flows through water or glass. However, it cannot explain the action that takes place when light strikes certain types of semiconductor materials, and it is this resultant action that forms the basis for much of the optoelectronic theory presented in this section. In order to explain why and how semiconductor materials are affected by light it is necessary to assume that light has additional characteristics.

To adequately explain the operation of the optoelectronic devices included in this section it is necessary to consider an additional aspect of light as explained by basic **quantum** theory. Quantum theory acknowledges that light has wave-like characteristics but it also states that a light wave behaves as if it consisted of many tiny particles. Each of these tiny particles represents a discrete quanta or packet of energy and is called a **photon**.

The photons within a light wave are unchanged particles and their energy content is determined by the frequency of the wave. The higher the frequency, the more energy each photon will contain. This means that the light waves at the upper end of the light spectrum possess more energy than the ones at the lower end of the spectrum. This same rule also applies to other types of electromagnetic radiation. For example, X-rays have a higher energy content than light waves while light waves possess more energy than radio waves.

Therefore, light has a dual personality. It propagates through space like radio waves, but it behaves as if it contains many tiny particles. This particle-like aspect of light will be used in this section to explain the action that takes place in various types of optoelectronic components.

Light Sensitive Devices

Light-sensitive devices respond to changes in light intensity by changing their internal resistance or by generating an output voltage. We will now examine some of the most important of these devices.

PHOTOCONDUCTIVE CELLS

The **photoconductive cell** is one of the oldest optoelectronic components. It is nothing more than a light-sensitive resistor whose internal resistance changes with light intensity. The resistance of the device decreases nonlinearly with an increase in light intensity. In other words the resistance decreases, but the decrease is not exactly proportional to the increase in light.

Photoconductive cells are usually made from light-sensitive materials such as cadmium sulfide (Cd S) or cadmium selenide (Cd Se), although other materials such as lead sulfide and lead telluride have been used. These materials may also be doped with other materials such as copper or chlorine to control the exact manner in which the resistance of the device varies with light intensity.

Figure 3-40 shows how a typical photoconductive cell is constructed. A thin layer of light-sensitive material is formed on an insulating substrate which is usually made from glass or ceramic materials. Then two metal electrodes are deposited on the light-sensitive material as thin layers. The top view (Figure 3-40A) shows that the electrodes do not touch but leave an S-shaped portion of the light-sensitive material exposed. This allows greater contact length but at the same time confines the light-sensitive material to a relatively small area between the electrodes. Two leads are also inserted through the substrate and soldered to the electrodes as shown in the side view in Figure 3-40B. The photoconductive cell is often mounted in a metal or plastic case (not shown) which has a glass window that will allow light to strike the light-sensitive material. Also, the electrodes used with some cells may be arranged in more complicated patterns and the entire cells may be quite large (1 or more inches in diameter) or relatively small (less than 0.25 inches in diameter).

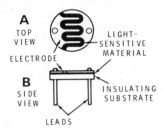

Figure 3-40
A typical photoconductive cell.
(A) topview,
(B) sideview.

Photoconductive cells are more sensitive to light than other types of light-sensitive devices. The resistance of a typical cell might be as high as several hundred megohms when the light striking its surface (its illumination) is zero (complete darkness) and as low as several hundred ohms when the illumination is over 9 footcandles. This represents a tremendous change in resistance for a relatively small change in illumination. This extreme sensitivity makes the photoconductive cell suitable for applications where light levels are low and where the changes in light intensity are small. However, these devices do have certain disadvantages. Their greatest disadvantage is the fact that they respond slowly to changes in illumination. In fact they have the slowest response of all light-sensitive devices. Also, they have a light memory or history effect. In other words, when the light level changes, the cell tends to remember previous illumination. The resistance of the cell at a specific light level is a function of the intensity, the duration of its previous exposure, and the length of time since that exposure.

Most photoconductive cells can withstand relatively high operating voltages. Typical devices will have maximum voltage ratings of 100, 200, or 300 volts DC. However, the maximum power consumption for these devices is relatively low. Maximum power ratings of 30 milliwatts to 300 milliwatts are typical.

The photoconductive cell is often represented by one of the schematic symbols shown in Figure 3-41. The symbol in Figure 3-41A consists of a resistor symbol inside of a circle. Two arrows are also used to show that the device is light-sensitive. The symbol in Figure 3-41B is similar but it contains the Greek letter λ (lambda) which is commonly used to represent the wavelength of light.

Figure 3-41
Commonly used photoconductive cell schematic symbols.

Photoconductive cells have many applications in electronics. For example, they are often used in devices such as intrusion detectors and automatic door openers where it is necessary to sense the presence or absence of light. However, they may also be used in precision test instruments which can measure the intensity of light. A simple intrusion detector circuit is shown in Figure 3-42. The light source projects a narrow beam of light onto the cell and this causes the cell to exhibit a relatively low resistance. The cell is in series with a sensitive AC relay and its 120 volt AC, 60 Hz power source. The cell allows sufficient current to flow through the circuit and energize the relay. When an intruder breaks the light beam, the cell's resistance increases considerably and the relay is deactivated. At this time the appropriate relay contacts close and apply power (from a separate DC source) to an alarm, which sounds a warning. A relay is used because it is capable of controlling the relatively high current that is needed to operate the alarm. When a large relay is used, the photocontuctive cell (because of its low power or current rating) may not be able to directly control the relay. In such a case the photoconductive cell is used to control a suitable amplifier circuit which in turn generates enough current to drive the relay.

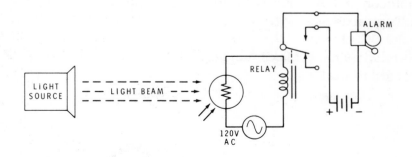

Figure 3-42
A basic intrusion detector circuit.

Since the photoconductive cell is constructed from a bulk material and does not have a PN junction, it is a bidirectional device. In other words it exhibits the same resistance in either direction and may therefore be used to control either DC or AC. Due to its bulk construction, the photoconductive cell is often referred to as a **bulk photoconductor**. However, you may also see it referred to as a **photoresistive cell** or simply a **photocell**.

PHOTOVOLTAIC CELLS

The **photovoltaic cell** is a device that converts light energy into electrical energy. When exposed to light this device generates a voltage across its terminals that increases with the light intensity. The photovoltaic cell has been used for a number of years in various military and space applications. It is commonly used aboard satellites and spacecraft to convert solar energy into electrical power which can be used to operate various types of electronic equipment. Since most of its applications generally involve the conversion of solar energy into electrical energy, this device is commonly referred to as a **solar cell**.

The photovoltaic cell is a junction device which is made from semiconductor materials. Although many different semiconductor materials have been used, the device is usually made from silicon. The structure of a silicon photovoltaic cell is shown in Figure 3-43. The device has a P-type layer and an N-type layer which form a PN junction. A metal backplate or support is placed against the N-type layer. Also, a metal ring is attached to the outer edge of the P-type layer. These pieces of metal serve as electrical contacts to which external leads may be attached.

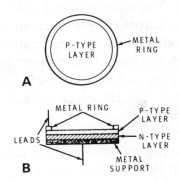

Figure 3-43
A basic silicon photovoltaic cell.
(A) topview,
(B) sideview.

The photovoltaic cell has a large surface area which can collect as much light as possible. Light strikes the top semiconductor layer within the metal ring.

Since the photovoltaic cell has a PN junction, a depletion region (an area void of majority carriers) forms in the vicinity of the junction. If the cell was forward-biased like a conventional PN junction diode, the free electrons and holes in the device would be forced to combine at the junction and forward current would flow. However, the photovoltaic cell is not used in this manner. Instead of responding to an external voltage, the device actually generates a voltage in response to light energy which strikes its surface.

In order to generate a voltage, the top layer of the photovoltaic cell must be exposed to light. The light energy striking the cell is assumed to consist of many tiny particles or photons. These photons are absorbed by the semiconductor material. A photon can intercept an atom within the semiconductor material and impart much of its energy to the atom. If sufficient energy is added to the atom, a valence electron may be knocked out of its orbit and become a free electron. This will leave the atom positively charged and a hole will be left behind at the valence site. In other words a photon can produce an electron-hole pair. This electron and hole can drift through the semiconductor material. Additional electron-hole pairs are also produced by other photons.

Some of the free electrons and holes generated by the light energy are produced within the depletion region while others are generated outside of the region but are drawn into it. The free electrons in the region are swept from the P-type to the N-type material and the holes are drawn in the opposite direction. The electrons and holes flowing in this manner produce a small voltage across the PN junction, and if a load resistance is connected across the cell's leads, this internal voltage will cause a small current to flow through the load. This current will flow from the N-type material, through the load and back to the P-type material thus making the N and P regions act like the negative and positive terminals of a battery.

All of the photons striking the photovoltaic cell do not create electron-hole pairs and many of the electrons and holes which separate to form pairs eventually recombine. The cell is therefore a highly inefficient device for converting light energy into electrical power. When this efficiency is expressed in terms of electrical power output compared to the total power contained in the input light energy, most cells will have efficiencies that range from 3 percent up to a maximum of 15 percent.

As you might expect, the output voltage produced by a photovoltaic cell is quite low. These devices usually require high light levels in order to provide useful output power. Typical applications require an illumination of at least 500 to 1000 footcandles. At 2000 footcandles, the average open-circuit (no load connected) output voltage of a typical cell is approximately 0.45 volts. When loaded, a typical cell may provide as much as 50 or 60 milliamperes of output load current. However, by connecting a large number of cells in series or parallel, any desired voltage rating or current capability can be obtained.

When used on spacecraft or satellites many photovoltaic cells are connected together, as explained above, to obtain sufficient power to operate electronic equipment or charge batteries. However, these devices are also used as individual components in various types of test instruments and equipment. For example, they are used in portable photographic light meters (which do not require batteries for operation). They are also used in movie projectors to detect a light beam which is projected through the film. The light beam is modulated (controlled) by a pattern or sound track that is printed near the edge of the film. In this way, the intensity of the light beam is made to vary according to the sounds (voice and music) that occur. The photovoltaic cell simply responds to the light fluctuations and produces a corresponding output voltage which can be further amplified and used to drive a loudspeaker which will convert the electrical energy back into sound. This application is shown in Figure 3-44.

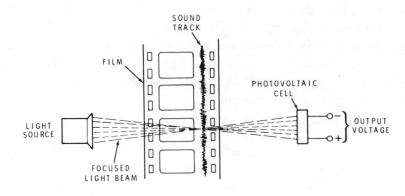

Figure 3-44
A photovoltaic cell used in a movie projector
sound reproducing system.

A schematic symbol that is commonly used to represent the photovoltaic cell is shown in Figure 3-45. This symbol indicates that the device is equivalent to a one-cell voltage source and the positive terminal of the device is identified by a plus (+) sign.

Figure 3-45
Commonly used schematic symbol for a photovoltaic cell.

PHOTODIODES

The photodiode is another light-sensitive device which utilizes a PN junction. It is constructed in a manner similar to the photovoltaic cell just described, but it is used in basically the same way as the photoconductive cell described earlier. In other words it is used essentially as a light-variable resistor.

The photodiode is a semiconductor device (usually made from silicon) and may be constructed in basically two ways. One type of photodiode utilizes a simple PN junction as shown in Figure 3-46A. A P-type region is formed on an N-type substrate as shown. This takes place through a round window that is etched into a silicon dioxide layer that is formed on top of the N-type substrate. Then a metal ring or window is formed over the silicon dioxide layer as shown. This window makes electrical contact with the P-type region and serves as an electrode to which an external lead can be attached. However, the window also accurately controls the area that will receive or respond to light. A metal base is then formed on the bottom N-type layer. This metal layer serves as a second electrode to which another lead is attached.

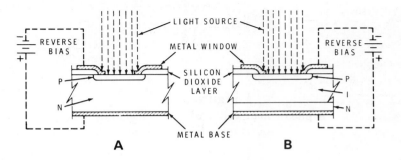

Figure 3-46
Basic construction of typical photodiodes.

The PN junction photodiode shown in Figure 3-46A operates on the same basic principle as the photovoltaic cell previously described. In fact, the photodiode may be used in basically the same manner as a photovoltaic cell. When used as a photovoltaic cell, the device is said to be operating in the **photovoltaic mode** and it will generate an output voltage (across its electrodes) that varies with the intensity of the light striking its top P-type layer. However, the photodiode is most commonly subjected to a reverse bias voltage as shown in Figure 3-46A. In other words its P-type region is made negative with respect to its N-type region. Under these conditions a wide depletion region forms around the PN junction. When photons enter this region to create electron-hole pairs, the separated electrons and holes are pulled in opposite directions because of the influence of the charges that exist on each side of the junction and the applied reverse bias. The electrons are drawn toward the positive side of the bias source (the N-type region) and the holes are attracted toward the negative side of the bias voltage (the P-type region). The separated electrons and holes therefore support a small current flow in the reverse direction through the photodiode. As the light intensity increases, more photons produce more electron-hole pairs which further increase the conductivity of the photodiode resulting in a proportionally higher current. When a photodiode is used in this manner it is said to be operating in the **photoconductive** or **photocurrent mode**.

The photodiode may also be constructed as shown in Figure 3-46B. This type of photodiode is similar to the type just described but there is one important difference. This device has an intrinsic (I) layer between its P and N regions and is commonly referred to as a **PIN** photodiode. The **intrinsic** layer has a very high resistance (a low conductivity) because it contains very few impurities. A depletion region will extend further into this I region than it would in a heavily doped semiconductor. The addition of the I layer results in a much wider depletion region for a given reverse bias voltage. This wider depletion area makes the PIN photodiode respond better to the lower light frequencies (longer wavelengths). The lower frequency photons have less energy content and tend to penetrate deeper into the diode's structure before producing electron-hole pairs and in many cases do not produce pairs. The wider depletion region in the PIN photodiode increases the chance that electron-hole pairs will be produced. The PIN photodiode is therefore more efficient over a wider range of light frequencies. The PIN device also has a lower internal capacitance due to the wide I region which acts like a wide dielectric between the P and N regions. This lower internal capacitance allows the device to respond faster to changes in light intensity. The wide depletion region also allows this device to provide a more linear change in reverse current for a given change in light intensity.

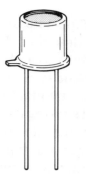

Figure 3-47
A typical photodiode package.

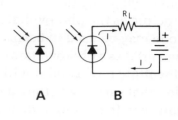

Figure 3-48
A photodiode symbol (A)
and a properly biased photodiode (B).

PN junction and PIN photodiodes are often mounted on an insulative platform or substrate and sealed within a metal case as shown in Figure 3-47. A glass window is provided at the top of the case, as shown, to allow light to enter and strike the photodiode. The two leads extend through the insulative base at the bottom of the case and are internally bonded (with fine wires) to the photodiode's electrodes.

Photodiodes have an important advantage over the photoconductive devices described earlier. A photodiode can respond much faster to changes in light intensity. In fact, the photodiode operates faster than any other type of photosensitive device. It is therefore useful in those applications where light fluctuates or changes intensity at a rapid rate. The major disadvantage with the photodiode is that its output photocurrent is relatively low when compared to other photoconductive devices.

Photodiodes and PIN photodiodes are both commonly represented by the same schematic symbol and several symbols have been used to represent these devices. A commonly used symbol is shown in Figure 3-48A. Notice that a conventional diode symbol is used with two arrows. The arrows point toward the diode to show that it responds to light. Figure 3-48B shows a properly biased photodiode. A load resistor (R_L) is also connected in series with the diode. The load resistor simply represents any resistive load which might be controlled by the photodiode as it varies its conductivity in accordance with input light intensity. The changes in the diode's conduction will cause the photocurrent (I) in the circuit to vary.

PHOTOTRANSISTORS

The phototransistor is also a PN junction device. However, it has two junctions instead of one like the photodiode just described. The phototransistor is constructed in a manner similar to an ordinary transistor, but this device is used in basically the same way as a photodiode.

The phototransistor is often constructed as shown in Figure 3-49. The process begins by taking an N-type substrate (usually silicon), which ultimately serves as the transistor's collector, and forming into this substrate a P-type region which serves as the base. Then an N-type region is formed into the P-type region to form the emitter. The phototransistor therefore resembles a standard NPN bipolar transistor in appearance. The device is often packaged much like the photodiode shown in Figure 3-47. However, in the case of the phototransistor, three leads are generally provided which connect to the emitter, base, and collector regions of the device. Also, the phototransistor is physically mounted under a transparent window so that light can strike its upper surface as shown in Figure 3-49.

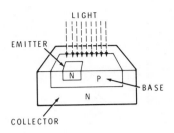

Figure 3-49
The construction of
a typical phototransistor.

The operation of a phototransistor is easier to understand if it is represented by the equivalent circuit shown in Figure 3-50. Notice that the circuit shown contains a photodiode which is connected across the base and collector of a conventional NPN bipolar transistor. If the equivalent circuit is biased by an external voltage source as shown, current will flow into the emitter lead of the circuit and out of the collector lead. The amount of current flowing through the circuit is controlled by the transistor in the equivalent circuit. This transistor conducts more or less depending on the conduction of the photodiode which in turn conducts more or less as the light striking it increases or decreases in intensity. If light intensity increases, the diode conducts more photocurrent (its resistance decreases) thus allowing more emitter-to-base current (commonly referred to as base current) to flow through the transistor. This increase in base current is relatively small but due to the transistor's amplifying ability this small base current is used to control the much larger emitter-to-collector current (also called collector current) flowing through this device. The increase in input light intensity causes a substantial increase in collector current. A decrease in light intensity would correspondingly cause a decrease in collector current.

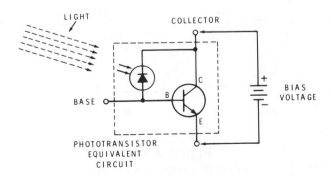

Figure 3-50
Equivalent circuit for a phototransistor.

Although the phototransistor has a base lead as well as emitter and collector leads, the base lead is used in very few applications. However, when the base is used, it is simply subjected to a bias voltage which will set the transistor's collector current to a specific value under a given set of conditions. In other words, the base may be used to adjust the phototransistor operating point. In most applications only the emitter and collector leads are used and the device is considered to have only two terminals.

An important difference between the phototransistor and the photodiode is in the amount of current that each device can handle. The phototransistor can produce much higher output current than a photodiode for a given light intensity because the phototransistor has a built-in amplifying ability. The phototransistor's higher sensitivity makes it useful for a wider range of applications than a photodiode. Unfortunately this higher sensitivity is offset by one important disadvantage. The phototransistor does not respond as quickly to changes in light intensity and therefore is not suitable for applications where an extremely fast response is required. Like other types of photosensitive devices the phototransistor is used in conjunction with a light source to perform many useful functions. It can be used in place of photoconductive cells and photodiodes in many applications and can provide an improvement in operation. Phototransistors are widely used in such applications as tachometers, photographic exposure controls, smoke and flame detectors, object counting, and mechanical positioning and moving systems.

A phototransistor is often represented by the symbol shown in Figure 3-51A and it is usually biased as shown in Figure 3-51B. As shown the phototransistor is used to control the current flowing through a load much like the photodiode shown in Figure 3-48B.

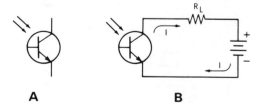

A   **B**

Figure 3-51
A phototransistor symbol (A)
and a properly biased transistor (B).

Light-Emitting Devices

Light-emitting devices are components which produce light when they are subjected to an electrical current or voltage. In other words, they simply convert electrical energy into light energy. For many years the incandescent lamp and the neon lamp were the most popular sources of light in various electrical applications. The incandescent lamp simply uses a metal filament which is placed inside of a glass bulb and air is drawn out of the bulb to produce a vacuum. Current flows through the filament causing it to heat up and produce light. The neon lamp utilizes two electrodes which are placed within a neon gas filled bulb. In this case current is made to flow from one electrode to the other through the gas and the gas ionizes and emits light.

The incandescent lamp produces a considerable amount of light, but its life expectancy is quite short. A typical lamp might last as long as 5000 hours. In addition to having a short life, the incandescent lamp responds slowly to changes in input electrical power. The incandescent lamp was (and still is) suitable for use as an indicator or for simply providing illumination, but due to its slow response it will not faithfully vary its light intensity in accordance with rapidly charging alternating currents. The incandescent lamp therefore cannot be effectively used to convert high frequency electrical signals (much above the audio range) into light energy which is suitable for transmission through space. The light energy produced by the incandescent lamp is not useful for carrying information which could subsequently be recovered or converted back into an electrical signal by a suitable light sensitive device.

The neon lamp has a somewhat longer life expectancy (typically 10,000 hours) than an incandescent lamp and a somewhat faster response to changes in input current. However, its output light intensity is much lower than that of an incandescent device. The neon lamp has been used for many years as an indicator or warning light and in certain applications to transmit low frequency AC signals or information in the form of light over very short distances. The neon lamp cannot be used simply for the purpose of providing illumination.

With all of their shortcomings, incandescent and neon lamps were used for many years simply because nothing better was available. However, in recent years a new type of light-emitting device was developed which has revolutionized the optoelectronics field. This newer device is a solid-state component, and it is physically stronger than the glass encased incandescent and neon devices. Like all semiconductor devices, it has an unlimited life expectancy. This new light-emitting device is referred to as a light-emitting diode or LED. Since it is such an important solid-state component we will examine the operation and construction of an LED in detail. Then we will see how it is used in various applications.

LED OPERATION

We have seen how light energy (photons) striking a PN junction diode can impart enough energy to the atoms within the device to produce electron-hole pairs. When the diode is reverse-biased these separated electrons and holes are swept across the diode's junction and support a small current through the device. However, the exact opposite is also possible. A PN junction diode can also emit light in response to an electric current. In this case, light energy (photons) is produced because electrons and holes are forced to recombine. When an electron and hole recombine, energy may be released in the form of a photon. The frequency (or wavelength) of the photons emitted in this manner is determined by the type of semiconductor material used in the construction of the diode.

The LED utilizes the principle just described. It is simply a PN junction diode that emits light through the recombination of electrons and holes when current is forced through its junction. The manner in which this occurs is illustrated in Figure 3-52. As shown in this figure, the LED must be forward-biased so that the negative terminal of the battery will inject electrons into the N-type layer (the cathode) and these electrons will move toward the junction. Corresponding holes will appear at the P-type or anode end of the diode (actually caused by the movement of electrons) and also appear to move toward the junction. The electrons and holes merge toward the junction where they may combine. If an electron possesses sufficient energy when it fills a hole it can produce a photon of light energy. Many such combinations can result in a substantial amount of light (many photons) being radiated from the device in various directions.

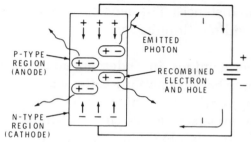

Figure 3-52
Basic operation of a light emitting diode.

At this time, you are probably wondering why the LED emits light and an ordinary diode does not. This is simply because most ordinary diodes are made from silicon and silicon is an opaque or impenetrable material as far as light energy is concerned. Any photons that are produced in an ordinary diode simply cannot escape. LED's are made from semiconductor materials that are semitransparent to light energy. Therefore, in an LED some of the light energy produced can escape from the device.

LED CONSTRUCTION

Many LED's are made of gallium arsenide (GaAs). The LED's made from this material emit light most efficiently at a wavelength of approximately 900 nanometers (10^{-9}) or 330,000 GHz which is in the infrared region of the light spectrum and is not visible to the human eye. Other materials are also used such as gallium-arsenide phosphide (GaAsP) which emits a visible red light at approximately 660 nanometers and gallium phosphide (GaP) which produces a visible green light at approximately 560 nanometers. The GaAsP device also offers a relatively wide range of possible output wavelengths by adjusting the amount of phosphide in the device. By adjusting the percentage of phosphide, the LED can be made to emit light at any wavelength between approximately 550 to 910 nanometers.

Although Figure 3-52 helps to illustrate the operation of an LED, it does not show how the device is constructed. The construction of a typical GaAsP LED is shown in Figure 3-53. Figure 3-53A shows a cross-section of the device and Figure 3-53B shows the entire LED chip. The construction begins with a gallium arsenide (GaAs) substrate. On this substrate a layer of gallium arsenide phosphide (GaAsP) is grown, however the concentration of gallium phosphide (GaP) in this layer is gradually increased from zero to the desired level. A gradual increase is required so that the crystalline structure of the substrate is not disturbed. During this growth period, an N-type impurity is added to make the layer an N-type material. The grown layer is then coated with a special insulative material and a window is etched into this insulator. A P-type impurity is then formed through the window into the layer and the PN junction is formed.

The P-type layer is made very thin so that the photons generated at or near the PN junction will have only a short distance to travel through the P-type layer and escape as shown in Figure 3-53A.

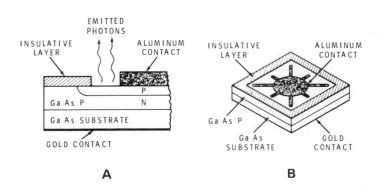

Figure 3-53
Basic construction of an LED.

The construction of the GaAsP LED is completed by attaching electrical contacts to the P-type region and the bottom of the substrate. The upper contact has a number of fingers extending outward so that current will be distributed evenly through the device when a forward bias voltage is applied across the contacts.

Once the LED is formed it must be mounted in a suitable package. Several types of packages are commonly used but all must fulfill one important requirement. All packages must be designed to optimize the emission of light from the LED. This factor is very important because the LED emits only a small amount of light. Therefore most packages contain a lens system which gathers and effectively magnifies the light produced by the LED. Also, various package shapes are used to obtain variations in the width of the emitted light beam or variations in the permissible viewing angle.

A typical LED package is shown in Figure 3-54. As shown, the package body and lens are one piece and are molded from plastic. The cathode and anode leads are inserted through the plastic case and extend up into the dome shaped top which serves as the lens. The bottom contact of the LED chip is attached directly to the cathode lead and the upper contact is connected to the anode lead by a thin wire which is bonded in place. The placement of the LED chip in this case is critical since the case serves as a lens which conducts light away from the LED and it also serves as a magnifier. In some cases the plastic lens will contain fine particles which help to diffuse the light or the entire case may be dyed or tinted with a color that enhances the natural light color emitted by the LED.

The LED package shown in Figure 3-54 is installed by simply pushing the lens through a suitable hole in a chassis or special bezel and snapping it in place. The leads are then soldered into place.

As you may suspect, the amount of light produced by an LED is small compared to an incandescent lamp. Most LED's generate a typical luminous intensity of only a few millicandelas, which is very low compared to even a miniature incandescent panel light which can produce many times that much light. However, LED's have several important advantages. First, they are extremely rugged. They also respond very quickly to changes in operating current and therefore can operate at extremely high speeds. They require very low operating voltages and are therefore compatible with integrated circuits, transistors, and other solid-state devices. They are relatively inexpensive when compared to incandescent devices. Also, they may be designed to emit a specific light color or narrow frequency range as compared to the incandescent lamp which emits a white light that contains a broad range of light frequencies.

The disadvantages associated with LED's (in addition to low light output) are similar to those which pertain to many types of solid-state components. They may be easily damaged by excessive voltage or current (beyond their maximum ratings) and their output radiant power is dependent on temperature.

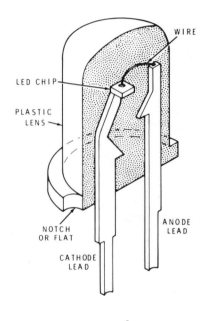

Figure 3-54
A typical LED package.

LED APPLICATIONS

In any application of an LED, the device is seldom used alone. The LED is usually connected in series with a resistor which limits the current flowing through the LED to the desired value. To operate the LED without this current limiting resistor would be risky since even a slight increase in operating voltage might cause an excessive amount of current to flow through the device. Some LED packages even contain built-in resistors (in chip form).

A schematic symbol that is commonly used to represent the LED is shown in Figure 3-55A. The correct way to bias an LED is shown in Figure 3-55B. The series resistor (R_S) must have a value which will limit the forward current (I_F) to the desired value based on the applied voltage (E) and the voltage drop across the LED which is approximately 1.6 volts. The following equation can be used to determine the required value of R_S:

$$R_S = \frac{E - 1.6}{I_F}$$

If we assume that E is equal to 6 volts and the I_F must equal 50 milliamperes (0.05 amperes) to obtain the desired light intensity, R_S must be equal to:

$$R_S = \frac{6 - 1.6}{0.05}$$

$$R_S = 88 \text{ ohms}$$

The previous equation is useful for determining R_S as long as the required R_S value is equal to or greater than 40 ohms.

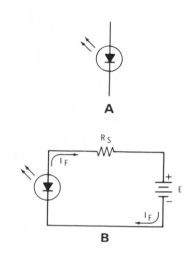

Figure 3-55
An LED symbol (A) and
a properly biased LED circuit (B).

Visible light producing LED's with their respective series resistors are often used as indicator lights to provide on and off indications. Individual LED's may even be arranged into specific patterns as shown in Figure 3-56. The LED's shown in this figure each illuminate one of seven segments arranged in a special pattern. The segments can be turned on or off to create the numbers 0 through 9 and certain letters. Such devices are referred to as 7 segment LED **displays**.

Figure 3-56
Typical LED numeric displays in a standard dual in line package. *Courtesy of Hewlett-Packard.*

LED's that emit infrared light may be used in intrusion detector systems if the light is properly focused and controlled. The infrared light cannot be seen by the human eye and is very effective in this application.

Infrared LED's are also commonly used in conjunction with light-sensitive devices, such as photodiodes or phototransistors, to form what is called an **optical coupler**. A typical optical coupler, which utilizes an LED and a phototransistor, is shown in Figure 3-57. The LED and photo-transistor chips are separated by a special type of light transmitting glass, and they are coupled only by the light beam produced by the LED. An electrical signal (varying current or voltage) applied to the LED's terminals (through two of the pins on the package) will produce changes in the light beam which in turn varies the conductivity of the phototransistor. When properly biased, the phototransistor will convert the varying light energy back into an electrical signal. This type of arrangement allows a signal to pass from one circuit to another but provides a high degree of electrical isolation between the circuits. Also, the LED responds quickly to input signal changes thus making it possible to transmit high frequency AC signals through the optical coupler.

Figure 3-57
A typical optical coupler which contains an LED and a
phototransistor. *Courtesy of General Electric.*

Liquid Crystals

Almost all of us have seen or, indeed, own a watch with a liquid crystal display (LCD). Not only are these devices visible in ambient light, they consume substantially less power than an LED display. This makes them ideal for watch displays. They are also used in many other devices such as the portable multimeter shown in Figure 3-58.

Figure 3-58
A portable multimeter that uses an LCD for low power consumption and high visibility.

A liquid crystal display is completely different from an LED display because the LED **generates** light, while the LCD **controls** light. That is, the LCD either blocks or permits the passage of light. Therefore, the LCD requires energy only to switch between these two states. On the other hand, the LED requires a continuous source of energy to emit light. Thus, the LCD consumes much less power than the LED.

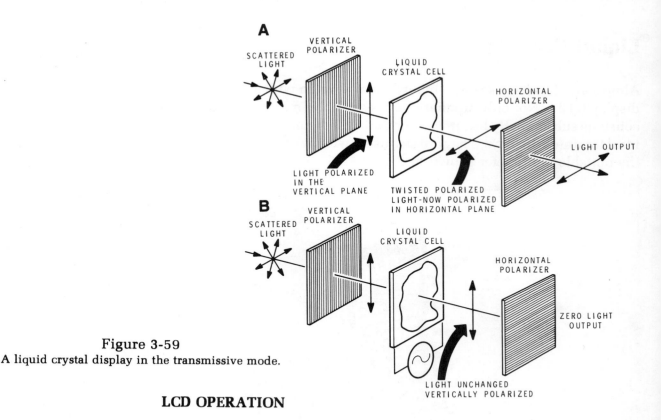

Figure 3-59
A liquid crystal display in the transmissive mode.

LCD OPERATION

LCDs have two basic modes of operation: the **transmissive** mode and the **reflective** mode.

An LCD in the transmissive mode is shown in Figure 3-59A. From the left of the diagram ambient light is traveling toward the LCD. The arrows represent the polarization of the light. Most ambient light is scattered light and, therefore, has waves that are polarized in almost all directions. However, the vertical polarizer of the LCD allows only vertically polarized light waves to pass and absorbs all other light waves. This vertically polarized light then passes through the liquid crystal. The properties of the liquid crystal are such that the light is twisted or rotated by 90°. Therefore, horizontally polarized light leaves the liquid crystal and passes uninhibited through the horizontal polarizer. Thus, the output will be some quantity of light, depending of course on the intensity of the ambient light.

However, if an AC voltage is applied across the liquid crystal, as shown in Figure 3-59B, the crystal molecules will alter their alignment in such a way that the light passing through will **not** be shifted 90° as before. Instead, it will pass through as vertically polarized light and be completely absorbed by the horizontal polarizer. Therefore, no light energy will pass to the output. Thus, the LCD can control light by either blocking it or permitting it to pass through.

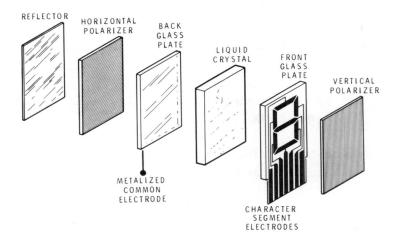

Figure 3-60
A reflective mode LCD.

An LCD operating in the reflective mode is shown in Figure 3-60. Here, with the addition of the reflector, the ambient light enters the front of the display, travels through the polarizers and liquid crystal, is reflected and then exits via the same path. In this LCD, a conductive film has been applied to the back glass plate to form a common electrode. This same conductive film is also applied to the front glass plate, except that it is applied in a special seven-segment character pattern as shown. Now, when voltage is applied to one or more of the character segments and to the common electrode, rotation of the light does not occur in the area of the activated segment(s). Therefore, the light from these segments is blocked at the horizontal polarizer. The other light passes on to the reflector and is reflected to the front of the display. The result is a dark pattern of activated character segments against a light background of reflected light.

You may have noticed that the common electrode and the character segment electrodes form the plates of a capacitor with the liquid crystal as a dielectric. Therefore, only an extremely small value of leakage current flows when a segment is activated. Thus, the drive requirements for the LCD are very low as is the power consumption.

Self-Review Questions

30. Light is considered to be _____ radiation that is visible to the human eye.

31. The light spectrum extends from _____ gigahertz to _____ gigahertz.

32. The invisible frequencies within the light spectrum are referred to as _____ and _____ rays.

33. Light has wave-like characteristics but it also behaves as if it consisted of many tiny particles known as _____ .

34. The photoconductive cell is basically a light-sensitive _____ .

35. The resistance of a photoconductive cell decreases as light intensity _____ .

36. The photovoltaic cell directly converts light into _____ energy.

37. The photovoltaic cell is sometimes referred to as a _____ cell.

38. The photodiode may operate in either the _____ or _____ mode.

39. A photodiode which has an intrinsic layer between its P and N regions is referred to as a _____ photodiode.

40. The phototransistor provides a higher output _____ than the photodiode.

41. The phototransistor's _____ and _____ leads are used in most applications.

42. Light-emitting devices convert electrical energy into _____ energy.

43. A properly biased LED emits light because of the recombination of _____ and _____ near its PN junction.

44. LED's made with gallium arsenide emit light in the _____ region.

45. Gallium phosphide LED's produce a visible _____ light.

46. An LED is to be operated at 40 mA with an applied voltage of 12 V. What value of series resistor must be used?

47. An optical coupler containing an LED and a photodiode will transmit information over a light beam but at the same time provide a high degree of electrical _____.

48. Identify the schematic symbols shown in Figure 3-61.

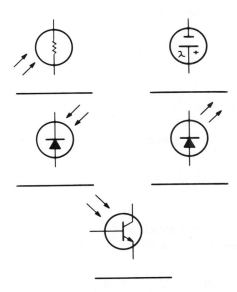

Figure 3-61
Fill in the blanks.

49. An LED display generates light, while a liquid crystal display _____ light.

50. What are the two basic modes of LCD operation? _____
_____

51. The primary advantage of the LCD over the LED display is its low
_____ .

Self-Review Answers

30. Light is considered to be **electro-magnetic** radiation that is visible to the human eye.

31. The light spectrum extends from **300** gigahertz to **300,000,000** gigahertz.

32. The invisible frequencies within the light spectrum are referred to as **infrared** and **ultraviolet** rays.

33. Light has wave-like characteristics but it also behaves as if it consisted of many tiny particles known as **photons**.

34. The photoconductive cell is basically a light sensitive **resistor**.

35. The resistance of a photoconductive cell decreases as light intensity **increases**.

36. The photovoltaic cell directly converts light into **electrical** energy.

37. The photovoltaic cell is sometimes referred to as a **solar** cell.

38. The photodiode may operated in either the **photovoltaic** or **photoconductive** mode.

39. A photodiode which has an intrinsic layer between its P and N regions is referred to as a **PIN** photodiode.

40. The phototransistor provides a higher output **current** than the photodiode.

41. The phototransistor's **collector** and **emitter** leads are used in most applications.

42. Light-emitting devices convert electrical energy into **light** energy.

43. A properly biased LED emits light because of the recombination of **electrons** and **holes** near its PN junction.

44. LED's made with **gallium arsenide** emit light in the **infrared** region.

45. Gallium phosphide LED's produce a visible **green** light.

46. $R_S = \dfrac{E - 1.6}{I_F}$

    $R_S = \dfrac{12 - 1.6}{0.04}$

    $R_S = 260 \; \Omega$

47. An optical coupler containing an LED and a photodiode will transmit information over a light beam but at the same time provide a high degree of electrical **isolation**.

48. See Figure 3-62.

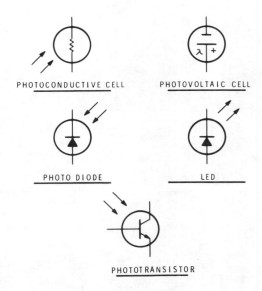

PHOTOCONDUCTIVE CELL

PHOTOVOLTAIC CELL

PHOTO DIODE

LED

PHOTOTRANSISTOR

Figure 3-62

49. An LED display generates light while a liquid crystal display **controls** light.

50. The two basic modes of LCD operation are **transmissive** and **reflective**.

51. The primary advantage of the LCD over the LED display is its low **power consumption**.

INTEGRATED CIRCUITS

In this section, you will be introduced to a type of solid-state device that is known as an integrated circuit or IC. The integrated circuit is actually a group of extremely small solid-state components which have been formed within or on a piece of semiconductor material and then connected to form a complete circuit. The IC is therefore a **solid-state circuit** and not an individual component like a diode or transistor.

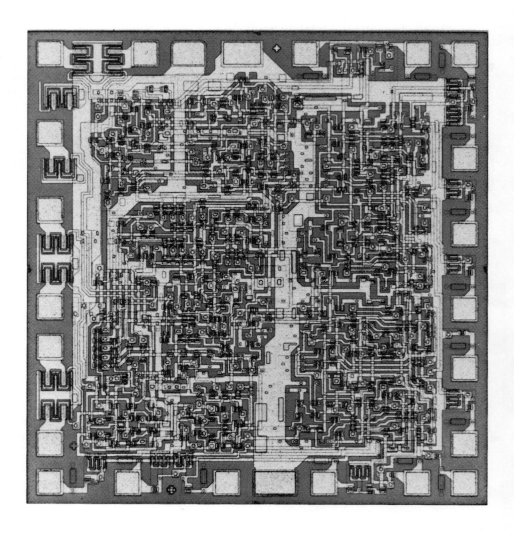

A magnified view of a single integrated circuit. *Courtesy of Motorola Inc.*

The Importance of IC's

Since their development in the late fifties, integrated circuits have had a tremendous effect on the electronics industry. Before IC's were developed, all electronic circuits were constructed with individual (discrete) components which were wired together. Various techniques were used to reduce the size of these discrete component circuits, but true miniaturization could not be obtained. The early vacuum tube circuits were quite large for the simple functions that they performed. The newer transistor circuits, although quite small and highly efficient when compared to their vacuum-tube counterparts, still did not offer the ultimate solution. It was the integrated circuit that finally made it possible to construct extremely small but highly efficient electronic circuits.

Typical miniature IC packages. *Courtesy Monsanto.*

Advantages and Disadvantages

The small size of the integrated circuit is its most apparent advantage. A typical IC can be constructed on a piece of semiconductor material that is less than one tenth of an inch square. Even after the IC has been packaged, it still occupies only a small amount of space.

The IC's small size also produces other fringe benefits. The smaller circuits consume less power than conventional circuits and therefore cost less to operate. They generate less heat and therefore generally do not require elaborate cooling or ventilation systems. The smaller circuits are also capable of operating at higher speeds because it takes less time for signals to travel through them. This is an important consideration in the computer field where thousands of decision-making circuits are used to provide rapid solutions to various types of problems.

The integrated circuit is also more reliable than a conventional circuit. This is because every component within the IC is a solid state device and because the components are permanently connected together. This eliminates any problems due to faulty connections. The IC also undergoes an extensive testing procedure which insures that it operates correctly.

Another potential advantage of IC's is low cost. However to take advantage of this, standard IC's must be used or very large quantities ordered. It can actually cost more to use IC's when a special purpose, one-of-a-kind circuit must be made.

The disadvantages of IC's are also related to their small size. Because of their size, IC's cannot handle high voltages or high power levels. High voltages can break down the insulation between the components within the IC because they are very close together. These components can also be easily damaged by the high heat levels associated with high power operation.

Another apparent disadvantage of IC's is that they cannot be repaired. However, this is offset by their low cost and the resultant easier troubleshooting. IC's are easier to troubleshoot because the problem need only be traced to a specific circuit instead of an individual component. This greatly simplifies the job of maintaining highly complex equipment.

Table I summarizes the advantages and disadvantages of integrated circuits.

ADVANTAGES	DISADVANTAGES
Small Size Consume Less Power High Speed Operation High Reliability Cheaper in Large Quantities	Cannot operate at High Power or High Voltage Cannot be repaired

Table I.
IC Advantages Vs. Disadvantages.

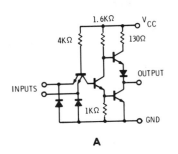

A

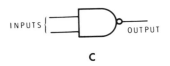

INPUTS		OUTPUT
0	0	1
0	1	1
1	0	1
1	1	0

B

INPUTS { OUTPUT

C

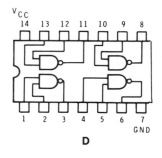

D

Figure 3-63
A typical TTL digital IC.

Applications of IC's

Integrated circuits may be placed into two general categories. They can be classified as either **digital** IC's or as **linear** IC's. Digital IC's are the most widely used devices. They are simply switching circuits which handle information and they are designed for use in various types of logic circuits and in digital computers. A linear IC provides an output signal that is proportional to the input signal applied to the device. Linear IC's are widely used to provide such functions as amplification and regulation. They are often used in television sets, FM receivers, electronic power supplies, and in various types of communications equipment.

It would be virtually impossible to consider all of the possible applications of digital and linear IC's, however, we can examine a few typical examples. First we will examine several digital IC's and then a typical linear IC.

Digital IC's

Most of the integrated circuits in use today are digital IC's. These devices are widely used in digital computers and portable electronic calculators to perform various arithmetic and decision making functions. Digital IC's are produced using both junction transistors and MOSFET construction techniques.

A typical digital IC which is formed by using junction transistor techniques is shown in Figure 3-63. A schematic diagram of the IC is shown in Figure 3-63A. Notice that only transistors, diodes, and resistors are used in the circuit and since it contains only 11 components, it is classified as a small scale integrated circuit. In this circuit, the transistors are the key elements and because of the unique manner in which they are connected, the circuit is commonly referred to as a **transistor-transistor logic (TTL)** circuit.

The TTL circuit in Figure 3-63A performs an important logic function. It is capable of comparing two input voltage levels, which must be equal to either 0 volts or approximately 3 volts, and provide an output voltage level (0 or 3 volts) depending on the input combination. The circuit performs what is commonly referred to as the NAND function and the circuit itself is referred to as a NAND gate since it provides a gating or switching action between two voltage levels. The NAND gate, like all digital circuits, is capable of recognizing only two voltage levels (sometimes called logic levels) at each of its inputs. Instead of referring to the specific voltages involved (which can vary with different types of digital circuits), it is common practice to refer to one level as a **high** logic level or a logic 1 and the other voltage level as a **low** logic level or a logic 0. A digital circuit can therefore be thought of as a device which responds to various high and low (1 or 0) logic levels and the actual voltages involved can be ignored. The table in Figure 3-63B shows the output levels (1 or 0) produced by the NAND gate when all possible combinations (only 4 are possible) of input levels have been applied. The NAND gate is therefore capable of making a simple decision based on the combinations of logic levels at its inputs and provide a specific output logic level for each combination.

Since many thousands of NAND gates are used in digital computers and other complex digital systems, it is not practical to draw the entire circuit each time it is shown on a schematic. Therefore, the NAND gate is usually represented by the symbol shown in Figure 3-63C. Notice that only the two inputs and the output of the circuit are represented.

It is common practice to construct not one but four of these NAND gate circuits on a single IC chip and mount the chip in a single package. The dual in-line package is widely used with IC's of this type. The outline of a typical dual in-line package is shown in Figure 3-63D. This package outline drawing shows how the various NAND gates are internally connected to the package leads. Notice that the IC package has 14 leads (also called pins), which are consecutively numbered in a counterclockwise direction. The package also has a notch at one end, which serves as a key to help locate pin number 1. Notice that pins 1 and 2 serve as inputs to one gate and pin 3 provides the output connection. Power is simultaneously applied to all four circuits through pins 14 and 7.

The package outline drawing shown in Figure 3-63D is typical of those provided by IC manufacturers in their specification sheets. They may also provide a schematic of the particular circuit involved and, of course, will always provide the important electrical characteristics of the circuit. In many cases, the circuit designer or engineer is more interested in what the circuit can do and is less interested in how it does it or how it is constructed. Therefore, specification sheets are likely to contain more mechanical and electrical information which pertains to the overall performance of the IC and very little information relating to its internal construction.

A typical example of a digital IC which is formed by using MOSFET techniques is shown in Figure 3-64. This IC utilizes one of the newest and most advanced construction techniques. It contains both P-channel and N-channel enhancement mode MOS field-effect transistors and is commonly referred to as a complementary-symmetry/metal-oxide semiconductor IC or simply a CMOS IC. CMOS circuits containing P- and N-channel MOSFET's are now widely used because they have many advantages over other types of digital circuits. They consume less power than other types of digital IC's and they have good temperature stability. They can operate over a wide range of supply voltages (typically 3 to 15 volts), as compared to TTL circuits which require an accurate 5 volt supply. CMOS circuits also have a high input resistance, which makes it possible to connect a large number of circuit inputs to a single output without loading down the output and disrupting circuit operation. This is an extremely important advantage in digital equipment where thousands of circuits are used.

The circuit shown in Figure 3-64A contains four MOSFET's which are interconnected so that they can perform a useful logic function. The resulting circuit is referred to as a NOR gate, and like the NAND gate previously mentioned, it is a fundamental building block that is used to construct complex digital circuits. However, the NOR gate responds differently to various combinations of input voltage levels. The output levels (1 or 0) produced by the NOR gate for all possible combinations of input levels are shown in the table of Figure 3-64B.

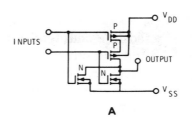

A

INPUTS		OUTPUT
0	0	1
0	1	0
1	0	0
1	1	0

B

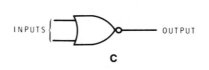

C

Figure 3-64
A CMOS digital IC.

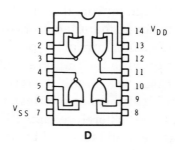

D

The NOR gate symbol is shown in Figure 3-64C. This symbol is generally used is place of the actual schematic. Four of these NOR gates are usually formed on one IC chip and mounted in a single IC package. An outline drawing of a typical dual in-line package, which contains four NOR gates, is shown in Figure 3-64D.

Both TTL and CMOS circuits can be used to perform NAND or NOR functions or a variety of other logic functions that must be performed in a highly complex digital system. The circuits shown in Figures 3-63 and 3-64 are therefore typical of the small scale integrated circuits used in digital equipment. These circuits may be thought of as basic building blocks which can be used to construct complex digital systems that can perform useful operations. Digital ICs and their applications will be discussed in detail in a later unit.

Linear IC's

As mentioned previously, linear circuits provide outputs that are proportional to their inputs. They do not switch between two states like digital circuits. The most popular linear circuits are the types that are designed to amplify DC and AC voltages. In fact, a high performance amplifier circuit, known as an **operational amplifier**, is widely used in various types of electronic equipment.

The operational amplifier can amplify DC or AC voltages and has an extremely high gain. An operational amplifier can be constructed with discrete components, but it is more commonly produced in IC form and therefore sold as a complete package which is designed to meet certain specifications. However, the operational amplifier is designed so that it can be used in a variety of applications. Its gain (ability to amplify) can be controlled by using additional external components and it usually has built-in features which make it possible to adjust its operation in various ways.

A typical operational amplifier circuit is shown in Figure 3-65A. The circuit contains transistors, resistors, and capacitors which are interconnected to form a highly efficient amplifying circuit. The circuit has two inputs and one output as shown. One input is commonly referred to as the plus (+) or non-inverting input and the other is referred to as the minus (−) or inverting input. The circuit will amplify either DC or AC signals applied to either input. However, signals applied to the (+) input are not inverted when they appear at the output. In other words, as the input voltage goes positive or negative, the output voltage correspondingly goes positive or negative. When a signal is applied to the (−) input, inversion takes place. In other words, the polarity of the output signal is always opposite to that of the input signal. This unique feature greatly increases the versatility of the circuit.

When the input voltage is equal to zero, the output voltage should also be equal to zero. However, in practice the output voltage may be offset by a slight amount since component tolerances make it impossible to construct a perfectly balanced circuit. Therefore, two offset null terminals are provided so that the circuit can be appropriately balanced. This is done by simply connecting the opposite ends of a potentiometer to the offset null terminals, but the arm of the potentiomenter is connected to circuit ground. The potentiometer may then be adjusted to balance the circuit.

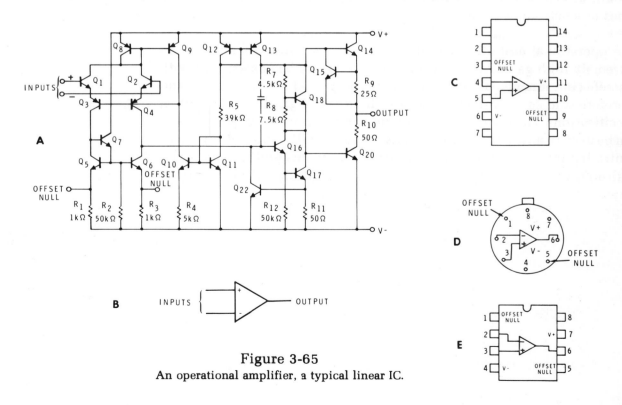

Figure 3-65
An operational amplifier, a typical linear IC.

Power is applied to the operational amplifier through terminals V+ and V−. The circuit therefore requires a positive voltage source and a negative voltage source. Most operational amplifiers can operate over a reasonably wide range of supply voltages, but these voltages should never exceed the maximum limits set by the manufacturer of the device. Operational amplifiers also consume very little power. Most units have maximum power dissipation ratings of 500 milliwatts or less.

The operational amplifier has an extremely high input resistance, but its output resistance is very low. The device also has an extremely high voltage gain. Many units are guaranteed to amplify an input voltage by at least 15,000 to 20,000 times and some of these units have typical gains that can exceed several hundred thousand or even more than a million.

In most cases, operational amplifiers are not used alone and their full amplifying capabilities are not utilized. Instead, it is common practice to connect external components to the amplifier in a manner which will allow a small portion of the output signal to return to the input and control the overall gain of the circuit. A lower gain is obtained in this manner, but circuit operation becomes more stable and predictable.

The operational amplifier is commonly represented by the symbol shown in Figure 3-65B. Notice that the (+) and (−) inputs are identified in the symbol. The operational amplifier circuit is generally packaged in a variety of ways to suit a broad range of applications. For example, the circuit in Figure 3-65A is available in a dual in-line package (DIP) as shown in Figure 3-65C or in a metal can packaged as shown in Figure 3-65D. It is even available as a mini DIP as shown in Figure 3-65E.

Operational amplifiers are used in various types of electronic equipment. They are the most important components used in electronic analog computers because their linear characteristics can be used to provide multiplying and summing operations. When used in analog computers, voltages are used to represent (are analogous to) the actual quantities to be multiplied or added. The extremely small operational amplifier IC's are suitable for use in portable electronic equipment where weight and power consumption must be held to a minimum. They are also suitable for use in portable test instruments and in communications equipment.

Various types of voltage regulator circuits are also constructed in IC form. These linear devices are used to convert an unregulated DC voltage (obtained through AC rectification) into a regulated DC output voltage which remains essentially constant while supplying a wide range of output currents. These voltage regulator IC's have replaced many of the discrete component regulators that were once widely used. Some IC regulators provide only one fixed output voltage, but other types are available which have adjustable outputs.

Special types of linear IC's are also designed for specific applications. For example, special linear IC's are designed for use in FM receivers where they are used to detect (recover the information in) FM signals. Some IC's are designed for use in solid-state color television receivers where they are used to detect, process and automatically control the chroma (color) signals. Others are used to provide extremely simple operations such as generating a signal to operate a lamp or some other indicator when an FM receiver or TV receiver is properly tuned. Some linear IC's are even used as an interface between digital circuits when digital information must be transmitted over a long transmission line. Such devices are commonly referred to as line drivers and receivers and they are used in digital systems even though they are basically linear devices.

Linear IC's and their applications will be discussed in greater detail in later units.

Self-Review Questions

52. What is an integrated circuit? _____
_____
_____

53. List the advantages and disadvantages of integrated circuits. ___
_____
_____

54. Name the two types of IC's.

    1. _____

    2. _____

55. Digital circuits deal with only _____ distinct logic levels.
These are referred to as a _____ logic level and a
_____ logic level.

56. The circuit shown in Figure 3-63A is known as a _____
digital logic circuit, while that shown in Figure 3-64A is a
_____ digital logic circuit.

57. A CMOS digital IC usually has a higher input _____
than a TTL digital IC.

58. A linear circuit provides an output that is _____
_____ to its input.

59. One of the most popular linear IC's is the _____
amplifier.

60. The input resistance of an operational amplifier is
_____ _____, the output resistance is
_____ _____, and the voltage gain is
_____ _____ .

Self-Review Answers

52. An integrated circuit is a solid-state circuit made up of a group of extremely small solid-state components which have been formed within or on a piece of semiconductor material.

53. Integrated circuit **advantages**:

   small size

   consume less power

   high speed operation

   high reliability

   cheaper in large quantities

   Integrated circuit **disadvantages**:

   cannot operate at high power or high voltage levels

   cannot be repaired.

54. The two types of IC's are:

   1. **Digital**

   2. **Linear**

55. Digital circuits deal with only **2** distinct logic levels. These are referred to as a **high** logic level and a **low** logic level.

56. The circuit shown in Figure 3-63A is known as a **TTL** digital logic circuit, while that shown in Figure 3-64A is a **CMOS** digital logic circuit.

57. A CMOS digital IC usually has a higher input **impedance** than a TTL digital IC.

58. A linear circuit provides an output that is **proportional** to its input.

59. One of the most popular linear IC's is the **operational** amplifier.

60. The input resistance of an operational amplifier is **very high**, the output resistance is **very low**, and the voltage gain is **very high**.

Unit 4

ELECTRONIC CIRCUITS

CONTENTS

INTRODUCTION

In this unit you will apply your knowledge of electronic devices and components. You will learn how these devices are used in power supplies, audio and radio frequency amplifiers, and oscillators. These circuits are used in all electronic equipment and are, therefore, vital to your understanding of electronics.

The "Unit Objectives," on the following page, state exactly what you are expected to learn from this unit. Study this list now and refer to it often as you study the text.

UNIT OBJECTIVES

When you have completed this unit, you should be able to:

1.   Identify half-wave, full-wave, and bridge rectifier circuits.

2.   State the ripple frequency and voltage polarity present at the output of a rectifier circuit.

3.   State the purpose of a power supply filter and identify some common circuits.

4.   Determine the peak inverse voltage a diode is subjected to in a rectifier circuit.

5.   State how you can obtain a higher PIV than is available from a single diode.

6.   Identify diode protection circuitry and state its operation.

7.   State why electrolytic capacitors are used in power supply filters.

8.   Define "capacitor working voltage rating," and state how you can increase it.

9.   Identify voltage doubler circuits.

10.   State the two purposes of a bleeder resistor.

11.   Identify zener diodes and IC voltage regulators.

12.   List the two types of audio amplifiers.

13. Find the gain or loss in decibels when given two power, voltage, or current levels.

14. Identify the four types of audio amplifier coupling circuits.

15. Define class A, B, AB, and C amplifier operation.

16. Find amplifier efficiency when given input and output power.

17. Name the basic operational amplifier terminals and state their function.

18. List the characteristics of an ideal operational amplifier.

19. Identify the three basic op amp circuit configurations and find the gain of each.

20. Identify the three types of coupling used in RF amplifiers.

21. State the requirements for an oscillator.

22. Identify the schematic diagrams of Armstrong. Hartley, Colpitts, and crystal oscillators.

23. Define piezoelectric effect.

24. State what determines the natural frequency of a crystal.

25. State the advantages and disadvantages of crystal oscillators.

POWER SUPPLIES

Most types of electronic equipment require direct current. In a few cases, such as transistor radios, this direct current is supplied by a battery. But more often, electronic devices are operated from the AC power line. In these cases, a special circuit, called a "power supply," converts the alternating current supplied by the power company to the direct current required by the electronic device. In this section you will study several circuits that are used for this purpose.

Rectifier Circuits

The heart of the power supply is the rectifier circuit, which converts the AC sine wave to a pulsating DC voltage. This is the first step in producing the smooth DC voltage required by the electronic circuits. Here, we will discuss three different types of rectifier circuits: the **half-wave, full-wave**, and **bridge** rectifier circuits.

HALF-WAVE RECTIFIER

The half-wave rectifier circuit is shown in Figure 4-1A. The load which requires the direct current is represented by a resistor. The solid-state diode connected in series with the load acts like a one-way switch that allows current to flow only in one direction.

The operation of the circuit during the positive half cycle is shown in Figure 4-1B. The anode of the diode is positive. Therefore, the diode conducts, allowing current to flow through the load. Thus, the positive half cycle of the input sine wave is developed across the load.

The next half cycle is shown in Figure 4-1C. Here, the anode is negative and the diode cannot conduct. Therefore, no current flows through the load and no voltage is developed across it.

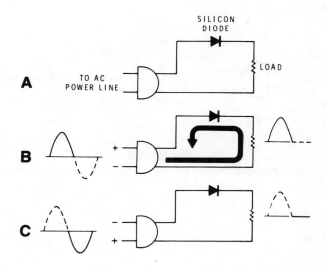

Figure 4-1
Half-wave rectifier circuit.

As you can see, the AC sine wave is changed to a pulsating DC voltage. This voltage is not suitable for most loads. However, this is only the first stage of the power supply. Later stages smooth out the ripples and change the pulsating DC to a steady DC.

The half-wave rectifier shown in Figure 4-1 will always produce the same output voltage level because it is connected across the 115 VAC line. To obtain higher or lower voltage levels, a step-up or step-down transformer is used. For example, Figure 4-2 shows a half-wave rectifier with a step-down transformer. Here the 115 VAC primary voltage is stepped down to 11.5 volts. It is then rectified and used to supply the load with approximately 11 VDC.

Ripple Frequency   The half-wave rectifier gets its name from the fact that it operates only during one half of each input cycle. Its output consists of a series of pulses. The frequency of these pulses is the same as the input frequency. Thus, when it is operated from the 60 Hz line, the frequency of the pulses will be 60 Hz. This is called the ripple frequency.

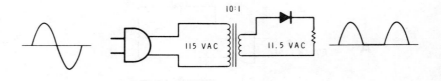

Figure 4-2
Half-wave rectifier circuit with a step-down transformer.

Output Polarity   A rectifier can produce either a negative or a positive output voltage. The output polarity depends on what point is connected to ground and which way the diode is connected. In Figure 4-3A, current can flow only in the direction indicated. This produces a positive voltage at the top of the load with respect to ground. Thus, this rectifier produces a positive output voltage.

If you need a negative output voltage, the diode can be turned around as shown in Figure 4-3B. Or, ground can be connected to a different point in the circuit as shown in Figure 4-3C. Either of these arrangements will produce a negative output voltage. However, if both changes are incorporated, the output voltage is positive again as shown in Figure 4-3D.

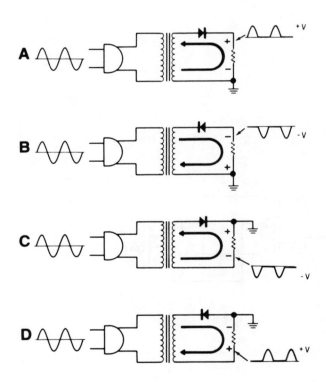

Figure 4-3
You can change the output polarity by turning the diode
around or by moving the ground connection.

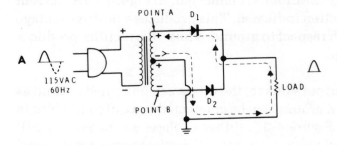

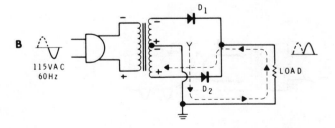

Figure 4-4
The full-wave rectifier.

FULL-WAVE RECTIFIER

Figure 4-4 shows a full-wave rectifier circuit. Notice that it uses two diodes and a center-tapped transformer. When the center-tap is grounded, the voltages at the opposite ends of the secondary are 180° out of phase with each other. Thus, when the voltage at point A swings positive with respect to ground, the voltage at point B swings negative. Let's examine the operation of the circuit during this half cycle.

Figure 4-4A shows that the anode of D_1 is positive while the anode of D_2 is negative. Therefore, only D_1 can conduct. Thus, current flows from the center tap up through the load and D_1 to the positive potential at the top of the secondary. The positive half cycle is therefore felt across the load.

Figure 4-4B shows what happens during the next half cycle, when the polarity of the voltage reverses. The anode of D_2 swings positive while the anode of D_1 swings negative. Therefore, D_1 cuts off and D_2 conducts. Thus, current flows from the center tap, through the load and D_2 to the positive potential at the bottom of the secondary.

Notice that current flows in the same direction through the load during **both** half cycles. This is a decided advantage over the half-wave rectifier, where current flows only during one half of the input cycle. Note also that there are two output pulses for every input cycle. Therefore, the ripple frequency is twice the input frequency. Thus, when it is operated from a 60 Hz line, the ripple frequency is 120 Hz.

Figure 4-5 illustrates a disadvantage of the full-wave rectifier. For a given transformer, it produces a peak output voltage which is half that of the half-wave rectifier. This is because the full-wave rectifier uses only one-half of the transformer secondary at one time.

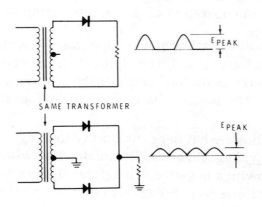

Figure 4-5
For a given transformer, E_{peak} is higher in the half-wave
rectifier.

BRIDGE RECTIFIER

The bridge rectifier circuit is shown in Figure 4-6. It consists of four diodes arranged so that current can flow through the load in only one direction. This circuit does not require a center-tapped transformer as the full-wave rectifier did. In fact it does not require a transformer at all except to provide a voltage step-up or step-down.

Figure 4-6A shows how current flows on the positive half cycle of the sine wave. Current flows from the bottom of the secondary through D_1, the load, and D_2, to the positive potential at the top of the secondary. With D_1 and D_2 acting as closed switches, the entire secondary voltage is developed across the load.

On the next half cycle, the polarities reverse, as shown in Figure 4-6B. The top of the secondary is now negative and the bottom is positive. Current flows from the top of the secondary through D_3, the load, and D_4 to the positive potential at the bottom of the secondary. Notice that the current flow through the load is always in the same direction. With D_3 and D_4 acting as closed switches, the entire secondary voltage is again developed across the load.

The advantages of the bridge rectifier are that it gives full-wave rectification and the full secondary voltage appears across the load.

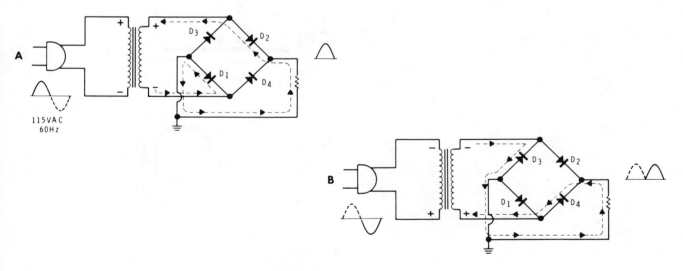

Figure 4-6
Current flow through the bridge rectifier.

Power Supply Filters

Most electronic circuits require a very smooth, constant voltage supply. Since the output of the rectifier is a pulsating DC voltage, it is unsuitable for nearly all electronic applications. Therefore, the rectifier circuit must be followed by a filter, which converts the pulsating DC into a smooth DC voltage.

CAPACITIVE FILTER

In its simplest form, the power supply filter may be nothing more than a capacitor connected across the output of the rectifier. This is shown in Figure 4-7A. Note that when the diode's anode is positive, current flows through the load. Simultaneously, a charge current flows to C_1, charging it to the peak of the input voltage.

When the diode's anode swings negative, as shown in Figure 4-7B, the current flow from the AC line is cut off. At this time, C_1 starts to discharge through the load. Before C_1 can completely discharge, the next cycle of the input AC starts, recharging C_1 to its peak voltage. Thus, C_1 supplies current to the load when the input AC is cut off.

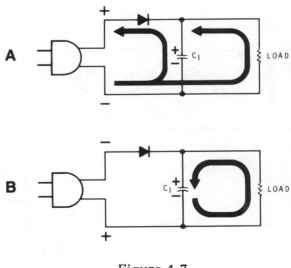

Figure 4-7
A capacitor power supply filter.

Figure 4-8A shows the output voltage waveform for an unfiltered half-wave rectifier. Figure 4-8B shows the output voltage when a small value capacitor is added. Note that the capacitor attempts to hold the voltage at a constant level. This is because of the basic nature of a capacitor to oppose changes in voltage, which it does by storing an electrical charge.

Figure 4-8C shows the output voltage when a much larger capacitor is used. The larger capacitor can store a greater electrical charge and, therefore, has a greater opposition to voltage changes.

Figure 4-8D shows the output of a full-wave or bridge rectifier with a capacitor filter. It is readily apparent that a full-wave rectified voltage is much easier to filter than a half-wave rectified voltage. This is due to the higher ripple frequency of the full-wave voltage.

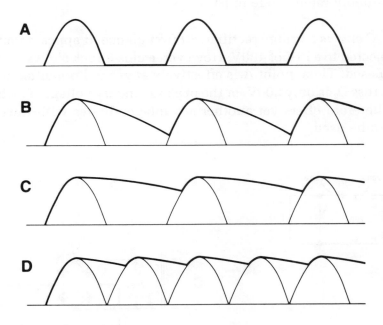

Figure 4-8

Rectifier output waveforms with varying degrees of capacitive filtering.

THE CAPACITOR'S EFFECT ON THE DIODES

Figure 4-9A shows a half-wave rectifier with a capacitor as a filter. During the half cycle in which the diode conducts, C_1 charges to the peak of the secondary voltage. Assume that this is +200 volts. The capacitor is large enough to hold the voltage across the load at approximately this level throughout the cycle.

During the next half cycle, the voltage across the secondary reverses. At the peak of the negative cycle, the anode of the diode is at −200 volts with respect to ground. At this point, the difference of potential across the diode is twice the peak value of the secondary, or 400 volts. Thus, the diode must have a peak inverse voltage (PIV) rating of at least 400V.

Figure 4-9B shows that a similar situation exists in the full-wave rectifier. With C_1 charged to the positive peak and the top of the secondary at the negative peak, D_1 experiences twice the peak secondary voltage. Here again, a suitably rated diode must be used.

Figure 4-9C shows a bridge rectifier. At first glance, it appears that D_2 is being subjected to a PIV of 400V. However, a closer look shows that D_3 is forward biased. Thus, point A is effectively at ground potential and the voltage across D_2 is only 200V, or the peak secondary voltage. The bridge rectifier therefore offers yet another advantage; diodes with lower PIV ratings can be used.

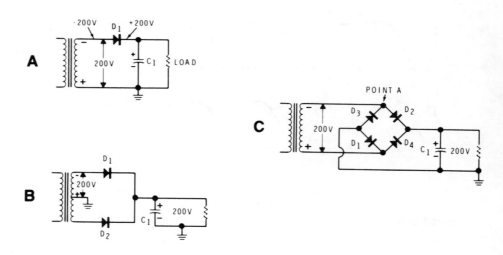

Figure 4-9
Determining the peak inverse voltage of the diodes.

When higher PIV ratings than a single diode can supply are required, several diodes can be connected in series. Then the inverse voltage will divide evenly between each diode, and the total PIV rating will be increased by the number of diodes in series.

As an example, Figure 4-10A shows three 1,000V PIV rated diodes connected in series. The PIV rating for this combination is 1,000V × 3 or 3,000V.

When you connect diodes in series, their reverse bias characteristics should match. That is, each diode's reverse "leakage" current should be the same. When this is not the case, unequal voltages will be dropped across the diodes and one diode may be subjected to a PIV greater than its maximum rating. To prevent this, **equalizing resistors** are connected in parallel with the diodes as shown in Figure 4-10B. Thus, the equalizing resistors force the reverse voltage drops to be equal.

Figure 4-10C shows an additional component which may be used on a diode "stack." The 0.01μF capacitors are used for "transient" protection. These transients, which are caused by DC switching at the load — or other sudden changes, are large voltage "spikes" that are often many times greater than the PIV rating of the diodes. These capacitors equalize and absorb any transients uniformly along the diode stack.

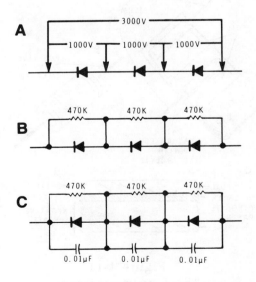

Figure 4-10
Diode protection circuitry.

ELECTROLYTIC CAPACITORS

An electrolytic capacitor is specially constructed to provide a large amount of capacitance in a small space. This makes it ideally suited for power supply circuits.

The construction of an electrolytic capacitor is shown in Figure 4-11. Sheets of metal foil are separated by a sheet of paper that is saturated with a conductive chemical paste called an electrolyte. The nonconductive dielectric is formed during the manufacturing process when a DC voltage is applied across the metal foil plates. As the DC currect flows, an extremely thin layer of aluminum oxide (see Detail 4-11C) builds up on the plate connected to the positive side of the DC voltage. The upper foil then becomes the positive plate of the capacitor. The oxide, because it is a good insulator, becomes the dielectric, and the electrolyte becomes the negative plate. Note that the capacitor is **polarized**. Figure 4-11B shows how the positive lead is marked. When this type of capacitor is connected in a circuit, you must connect the positive lead to the more positive voltage. (NOTE: On some electrolytic capacitors, only the **negative** lead may be marked.)

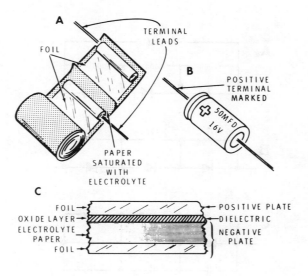

Figure 4-11
The electrolytic capacitor.

Because capacitance is inversely proportional to plate spacing, an electrolytic capacitor can have a very large capacitance due to its very thin dielectric. However, since the dieletric is so thin, a high voltage can cause the dielectric to arc over or break down. This is why capacitors have a **working voltage rating**. This is the maximum voltage the capacitor can withstand without the dielectric breaking down or arcing over.

When higher working voltage ratings are required, several capacitors can be connected in series as shown in Figure 4-12. Here, the maximum voltage rating is 450 V × 3 or 1350 V. Note the 30 kΩ "equalizing"resistors which must be used to equalize the voltage drops across each capacitor.

A "stack" of capacitors with equalizing resistors is shown in Figure 4-13. This is the power supply for a 2,000 watt amateur radio linear amplifier. The working voltage rating of the capacitor stack is 3,600 V!

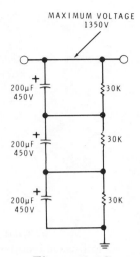

Figure 4-12
Using capacitors in series to increase the voltage rating.

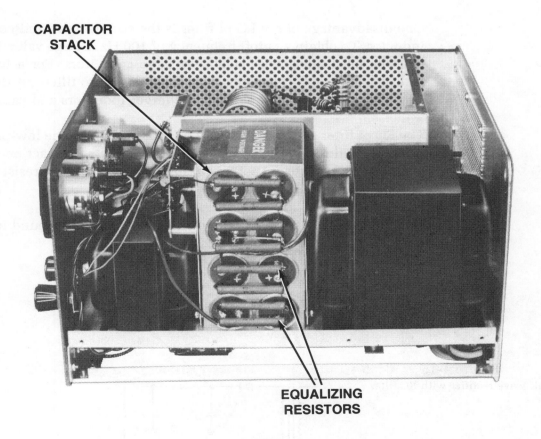

Figure 4-13
Capacitor stack in a high-power amplifier power supply.

PI FILTERS

Figure 4-14 shows a more elaborate type of power supply filter, the LC pi filter. This name is used because, in its schematic diagram, the circuit components are arranged to resemble the Greek letter pi (π). You may have recognized it as a low-pass filter. In this case, its cut off frequency is approximately 100 Hz. That is, it passes all frequencies below 100 Hz. Since the ripple frequency of the full-wave rectifier is 120 Hz, only DC is allowed to pass to the load.

Figure 4-14
Full-wave rectifier with LC filter.

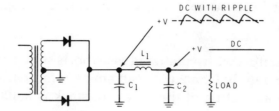

One disadvantage of the LC pi filter is the size, weight, and cost of the inductor. To obtain a cutoff frequency of 100 Hz, a large value inductor must be used along with relatively large capacitors. For a half-wave rectifier, the cutoff frequency must be 40-50 Hz to filter out the 60 Hz ripple component. This requires still larger inductors and capacitors.

The RC pi filter shown in Figure 4-15 works on the same low-pass filter principle and is a lot less expensive. However, its filter action is not as good as the LC filter. Another disadvantage is that the resistor drops voltage which could be used by the load.

Nonetheless, the RC filter is a good compromise and is found in a large number of power supplies.

Figure 4-15
Half-wave rectifier with RC filter.

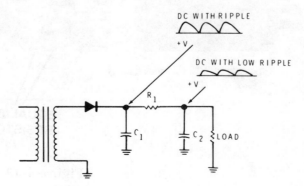

Voltage Doublers

Two rectifier circuits have been developed that can double the voltage from a transformer or the AC line. These are the half-wave and full-wave voltage doublers.

HALF-WAVE VOLTAGE DOUBLER

The half-wave voltage doubler, as shown in Figure 4-16A, consists of two diodes and two capacitors. It produces a DC output voltage which is approximately twice the peak value of the input AC sine wave. In this example, assume that the input is 115 VAC at 60 Hz.

Figure 4-16B shows how the circuit responds to the negative half cycle of the input sine wave. When the voltage at point A swings negative, D_1 conducts and current flows as shown by the arrows. This charges C_1, with the polarity shown, to the peak value of the input sine wave (about 162 volts). Since there is no immediate discharge path for C_1, the capacitor remains charged to this level until the next half cycle.

Figure 4-16C shows what happens on the positive half-cycle. At the peak of the cycle, point A is at + 162 V. Since C_1 is charged to 162 V, and the voltages are in series, they add: 162 V + 162 V = 324 V. Therefore, the voltage at point B is +324 V. This forward biases D_2 and current flows as shown, charging C_2 to +324 V. Thus, the load sees twice the peak value of the input AC, or 324 V.

Between the positive peaks of the sine waves, D_2 is cut off because C_2 holds its cathode at a high positive potential. During this period, C_2 discharges through the load. Thus, C_2 acts as a filter capacitor holding the voltage across the load fairly constant. Since C_2 is recharged only during the positive peak of the input sine wave, the ripple frequency is 60 Hz. Consequently, this circuit is called a half-wave voltage doubler.

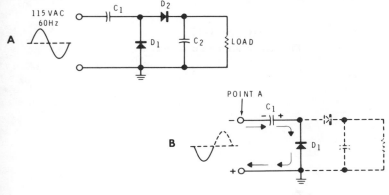

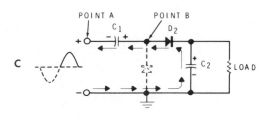

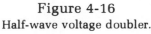

Figure 4-16
Half-wave voltage doubler.

FULL-WAVE VOLTAGE DOUBLER

The circuit for the full-wave voltage doubler is shown in Figure 4-17A. On the positive half cycle, C_1 charges through D_1 to the peak value of the AC input, as shown in Figure 4-17B. In this case C_1 charges to about 162 V.

On the negative half cycle, C_2 charges through D_2 as shown in Figure 4-17C. C_2 charges to the peak value of the AC input, or to about 162 V.

Once the capacitors are charged, D_1 and D_2 conduct only at the peaks of the AC input. Between the peaks, C_1 and C_2 discharge in series through the load as shown in Figure 4-17D. If each capacitor is charged to 162 V, then the total voltage across the load is about 324 V, or twice the peak value of the input AC. Here, the ripple frequency is 120 Hz, since the C_1-C_2 combination is recharged twice during each cycle.

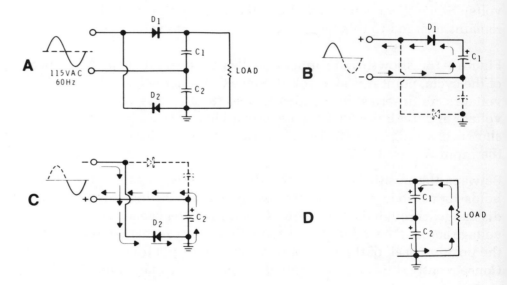

Figure 4-17
Full-wave voltage doubler.

Voltage Regulation

Ideally, the output of a power supply should be a constant voltage. Unfortunately, this is difficult to achieve. There are two factors which can cause the output voltage to change.

First, the AC line voltage is not constant. The so-called 115 VAC can vary from about 105 VAC to 125 VAC. This means that the peak AC voltage to which the rectifier responds can vary from about 148 V to 177 V. As you can see, the AC line voltage can cause as much as a 20% change in the DC output voltage.

The second thing that can change the DC output voltage is a change in the load resistance. In electronic equipment, the load can change as circuits are switched in and out. For example, in a TV Receiver, the load changes each time the video changes from white to black.

Variations in load resistance tend to change the applied DC voltage because the power supply filter capacitors are discharged more quickly. One way to decrease voltage variations is to use a bleeder resistor.

BLEEDER RESISTOR

Most power supplies use a bleeder resistor across the output, as shown in Figure 4-18. The bleeder serves two purposes. First, it discharges the filter capacitor when the power supply is turned off. This is a safety precaution, as filter capacitors can hold a charge for a very long time. Therefore, even though the power supply is turned off, they can present a very dangerous shock hazard. The bleeder eliminates this hazard by discharging the capacitors.

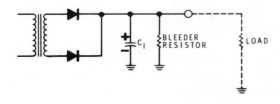

Figure 4-18
Using a bleeder resistor to improve voltage regulation.

The second purpose of the bleeder resistor is to reduce output voltage fluctuations. Figure 4-19 is a graph showing the effect of current drain on the power supply's output voltage. Note that when current is zero, the output voltage is 162 V, or the peak of the input AC. This is because the capacitors charge to the peak and hold it there. When current is drawn from the supply, the capacitors discharge and the output voltage drops rapidly, leveling off at about 120 V. This is a drop of 42 V, or a 35% change.

If a bleeder resistor that draws about 100 mA is used, the "no-load" output voltage will be approximately 130 V. Then, when full current is drawn from the supply, the voltage drops to 120 V. This is a drop of only 10 V, or an 8% change. Thus, the bleeder resistor has improved the power supply's **regulation**. That is, its no-load voltage is much closer to its full load voltage and output variations have been reduced.

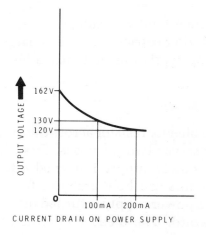

Figure 4-19
Effect of current drain on output voltage.

ZENER VOLTAGE REGULATOR

Ideally, the difference between no load and full load voltage would be zero. A circuit that comes close to accomplishing this ideal condition is known as a **voltage regulator**.

In an earlier unit, you studied the zener diode which, when reverse biased into breakdown, has a constant voltage drop. Thus, it can be used as a simple voltage regulator.

A zener diode regulator circuit is shown in Figure 4-20. Notice that the diode is connected in series with a resistor and that an unregulated DC input voltage is applied to these two components. The input voltage is connected so that the zener diode is reverse biased. The series resistor allows enough current to flow through the diode so that the device operates within its zener breakdown region.

In order for this circuit to function properly the input DC voltage must be higher than the zener breakdown voltage. The voltage across the diode will then be equal to the diode's zener voltage rating. The voltage across the resistor will be equal to the difference between the diode's zener voltage and the input DC voltage.

The input DC voltage is unregulated. This voltage will periodically increase above or decrease below its specified value. Therefore, it causes the current flowing through the zener diode and the series resistor to fluctuate. However, the diode is operating within its zener voltage region. Therefore, a wide range of current can flow through the diode while its zener voltage changes only slightly. Since the diode's voltage remains almost constant as the input voltage varies, the change in input voltage appears across the series resistor. Remember that these two components are in series and the sum of their voltage drops must always be equal to the input voltage.

The voltage across the zener diode is used as the output voltage for the regulator circuit. The output voltage is therefore equal to the diode's zener voltage. Since this voltage is held to a nearly constant value, it is referred to as a regulated voltage. You can change the output voltage of this regulator circuit by using a diode with a different zener voltage rating.

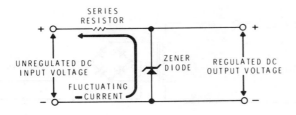

Figure 4-20
Basic zener regulator.

IC VOLTAGE REGULATORS

The trend in power supply design is toward integrated circuit (IC) regulators. These are available with a wide range of output voltages and currents. Generally, a single IC will contain a complex regulator circuit, and additional protection circuitry which makes it virtually blowout proof.

Some IC regulators are designed to deliver a fixed output voltage such as +5V, +12V, or −12V. However, others can be set up with external components to produce a range of voltages or even an adjustable voltage.

Some of the low voltage regulators can deliver 1A or more to the load without external components. Many regulators, though, are limited to 200 mA or below. If higher currents are required, additional components must be added externally.

Figure 4-21 shows a complete power supply that uses a 5V IC regulator. The regulator has only three terminals and it uses a transistor package. However, the IC contains over a dozen transistors arranged in a complex regulator circuit. Regulator IC's are therefore classified as linear IC's.

An unregulated DC voltage is supplied to the input terminal of the regulator. This voltage can be any value from about +10V to +35V. In the circuit shown, the bridge rectifier supplies +12.5V, which is filtered by C_1. The IC regulator accepts this unregulated input and produces a precise, regulated, +5V output. The circuit is protected by the 3/16 A slow-blow fuse, F1, as well as the IC's internal protection circuitry. Switch SW_1 is the on-off switch.

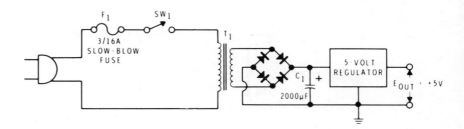

Figure 4-21
5-volt power supply using an IC voltage regulator.

Self-Review Questions

1. Identify the circuits shown in Figure 4-22.

    A. _Full wave rectifier_

    B. _Bridge rectifier._

    C. _Half wave rectifier_

2. What is the output ripple frequency and DC voltage polarity of the
    circuit shown in Figure 4-22A? _120 Hz_
    In Figure 4-22B?_120 Hz_
    In Figure 4-22C?_60 Hz_

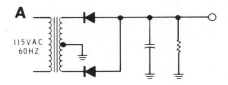

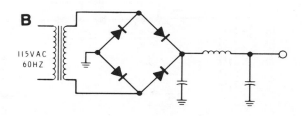

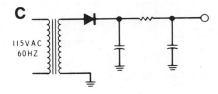

Figure 4-22
Identify these circuits.

3. What is the purpose of a power supply filter? _____
    converts pulsating DC to smooth DC

4.   What is the minimum PIV rating for D_1 in Figure 4-23? _____

400 v

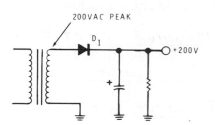

Figure 4-23
What is the PIV rating for D_1?

5.   Why are electrolytic capacitors used in power supply filters? __

Large capacitance in a small space

6.   Identify the circuit shown in Figure 4-24.

Full wave voltage doubler

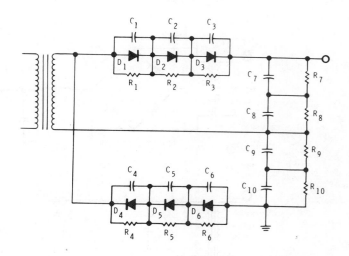

Figure 4-24
Identify this circuit.

7.   What is the purpose of connecting D_1, D_2 and D_3 in series in Figure 4-24? *To increase PIV rating*

8.   What is the purpose of C_1-C_3 and R_1-R_3 in Figure 4-24? _____

Diode protection

9. What is the purpose of connecting C₇ and C₈ in series in Figure 4-24? *Increase working voltage rating*

10. State the two purposes of a bleeder resistor. *1. Discharge capacitors 2. Improve voltage regulation*

11. Identify the circuits in Figure 4-25.

A. *IC voltage regulator*

B. *Zener voltage regulator*

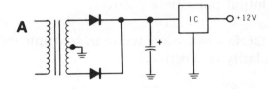

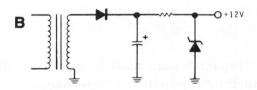

Figure 4-25
Identify these circuits.

Self-Review Answers

1. The circuits shown in Figure 4-22 are:

   A. Full-wave rectifier, capacitor filter.
   B. Bridge rectifier, LC pi filter.
   C. Half-wave rectifier, RC pi filter.

2. Figure 4-22A is a full-wave rectifier. Since the input is 60 Hz, the ripple frequency is 60 × 2 or 120 Hz. The output polarity is determined by the diode connections, in this case it is negative.

   Figure 4-22B is a bridge rectifier, so ripple frequency is 2 × 60 Hz = 120 Hz. Output polarity is positive.

   Figure 4-22C is a half-wave rectifier, so ripple frequency is 60 Hz. Output polarity is positive.

3. A power supply filter converts the pulsating DC from the rectifier into a smooth DC voltage.

4. Since Figure 4-23 has a capacitive filter, the PIV is twice the peak secondary voltage: 200 V × 2 = 400 V.

5. Electrolytic capacitors are used in power supply filters because they provide a large amount of capacitance in a small space.

6. The circuit shown in Figure 4-24 is a full-wave voltage doubler.

7. D_1, D_2, and D_3 are connected in series to increase the PIV rating.

8. The purpose of C_1-C_3 and R_1-R_3 in Figure 4-24 is diode protection. C_1-C_3 provide transient protection, while R_1-R_3 equalize the inverse voltage across the diode.

9. The purpose of connecting C_7 and C_8 in series is to increase the working voltage rating.

10. The purposes of a bleeder resistor are to discharge the filter capacitors when the power supply is turned off and to improve the voltage regulation of the supply.

11. The circuits shown in Figure 4-25 are :

A. IC voltage regulator.
B. Zener diode regulator.

AUDIO AMPLIFIERS

There are many applications where it is necessary to amplify AC signals that are within the audio frequency range, which extends from 20 Hz to 20,000 Hz. The circuits used to amplify these signals are generally referred to as audio frequency (AF) amplifiers, or simply audio amplifiers.

Voltage and Power Amplifiers

Audio amplifiers provide both voltage and power amplification, or **gain**. Many amplifiers are specifically designed to amplify low voltage audio signals and are classified as **voltage amplifiers**. In most cases, voltage amplifiers are used to increase the voltage level of an input signal to a value high enough to drive a power amplifier stage. The **power amplifier** can then supply a high output signal current to operate a loudspeaker or some other device requiring high power.

Amplifier Gain and the Decibel

As you learned in a previous unit, amplifier gain is the ratio of input to output. Therefore, amplifier power gain is:

$$A_p = \frac{P_{out}}{P_{in}}$$

Where A_p =   power gain

P_{out} =   output power

P_{in} =   input power

One of the most common ways to express gain or power ratio is with a unit of measurement called the **decibel**, which is abbreviated dB. Originally, the decibel was based on the response of the human ear to sound. At lower sound levels, the ear can detect relatively small changes. However, as the volume is increased, larger and larger sound level changes are required before the ear can detect any difference. The decibel is a unit of measurement for the smallest change that the ear can detect.

Starting with a barely discernable sound level, one milliwatt, for example, we will increase the volume in barely discernable steps and see, in Figure 4-26, how much power is required. Remember that in this case, our starting level, 0 dB, is equal to 1 milliwatt (1 mW).

A power change of approximately 26% is required before your ear can detect any change in sound level. Therefore, the first point where you would notice a change would be at the 1.26 mW level. And since 1 dB is the minimum discernable change, we have labeled this 1 on the horizontal axis.

Now, if you were to increase the sound level further until you could detect another change, it would require 26% more power than this 1.26 mW level, or approximately 1.6 mW. This is the 2 dB point.

At the 3 dB point, the sound level is up to 2 mW, or twice the original power. And as you look further up the curve on the graph, notice that a discernable (1 dB) change requires more and more power. This is called a **logarithmic** change. If you are not familiar with logarithms or would like a review, read "Appendix A."

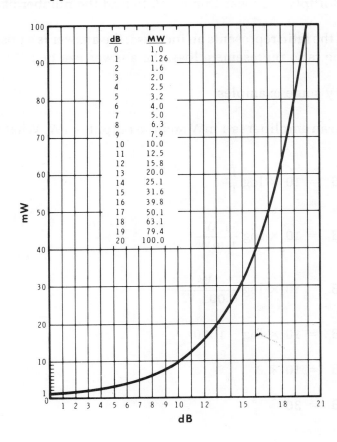

dB	MW
0	1.0
1	1.26
2	1.6
3	2.0
4	2.5
5	3.2
6	4.0
7	5.0
8	6.3
9	7.9
10	10.0
11	12.5
12	15.8
13	20.0
14	25.1
15	31.6
16	39.8
17	50.1
18	63.1
19	79.4
20	100.0

Figure 4-26
Decibels referenced to 1 milliwatt.

In our graph, you found that a power change from 1 mW to 2 mW was a 3 dB change. Notice also that a change in power from 10 mW to 20 mW is a change of 10 dB to 13 dB, a 3 dB change. Likewise, a change from 50 mW to 100 mW is a 3 dB change. Thus, you can see that dB is a measure of change, not a measure of actual power.

The formula for calculating decibels is:

$$dB = 10 \times \log \frac{P_2}{P_1}$$

To use the formula, follow this procedure:

1. Find the power ratio, making sure to use the same units for both P_1 and P_2. Always use the larger power level as P_2. Then the ratio will always be greater than 1, and you will avoid working with negative logarithms.

2. Find the logarithm of the power ratio.

3. Multiply the logarithm by 10 to find the number of decibels.

4. If the ratio represents an increase, the answer is a positive dB. If the ratio represents a loss, the answer is a negative dB.

Now let's try some examples.

The input to an amplifier is 1 mW and the output is 1W. What is the gain in dB?

$$dB = 10 \times \log \frac{P_2}{P_1}$$

$$dB = 10 \times \log \frac{1 \text{ W}}{1 \text{ mW}}$$

$$dB = 10 \times \log \frac{1 \text{W}}{0.001 \text{ W}}$$

$$dB = 10 \times \log \quad 1000$$

$$dB = 10 \times 3$$

$$dB = 30$$

Since there is an increase in power, the gain of the amplifier is $+30$ dB.

What is the gain or loss of a resistor network whose input is 100 mW and output is 10 mW?

$$dB = 10 \times \log \frac{100 \text{ mW}}{10 \text{ mW}}$$

$$dB = 10 \times \log \quad 10$$

$$dB = 10 \times 1$$

$$dB = 10$$

Since there is a decrease in power, the loss of the network is -10 dB.

It is also possible to express a voltage or current gain in dB. Since power is $\frac{E^2}{R}$, you can express the decibel formula as:

$$dB = 20 \times \log \frac{E_2}{E_1}$$

The two voltages must be across equal impedances or you cannot use this formula. The factor of 20 in the formula is because power is proportional to voltage squared, and doubling the logarithm corresponds to squaring the number.

As an example, for an increase from 1 mV to 1V, the dB voltage gain is:

$$dB = 20 \times \log \frac{E_2}{E_1}$$

$$dB = 20 \times \log \frac{1 \text{ V}}{0.001 \text{ V}}$$

$$dB = 20 \times \log \quad 1000$$

$$dB = 20 \times 3$$

$$dB = 60$$

The voltage gain is $+60$ dB.

The formula for current gain in dB is:

$$dB = 20 \times \log \frac{I_2}{I_1}$$

Note that the factor 20 also appears in this formula. This is because power is proportional to current squared.

A graph of voltage, current, and power ratios from 1 to 10 is shown in Figure 4-27. You can use this chart to convert power, voltage, or current gain to dB, or to convert dB to power, voltage, or current gain.

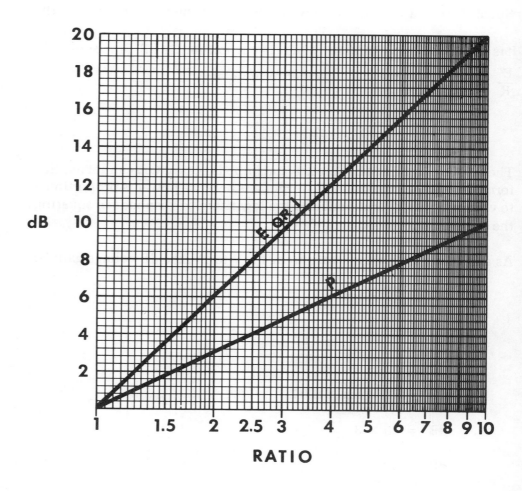

Figure 4-27
Current, voltage, and power ratios and their dB equivalents.

Coupling Circuits

In some applications, one amplifier stage (one transistor or tube) cannot provide enough amplification. Therefore, it becomes necessary to couple two or more amplifier stages together to obtain a higher overall gain.

When two amplifier stages are joined together, it must be done in a way which will not upset or disrupt the operation of either circuit. There are four basic coupling methods which are widely used. Let's examine each of these methods.

CAPACITIVE COUPLING

The capacitive, or RC, coupling technique is one of the most widely used methods. In this type of circuit, the signal must pass from one amplifier stage to the next through a coupling capacitor. The primary purpose of this coupling capacitor is to block DC voltage from the previous stage, and thus prevent bias changes. Figure 4-28 shows how transistor stages are capacitive coupled.

In order for the coupling capacitor to efficiently transfer the AC signal to the second stage, it must offer very little opposition to the AC signal current, even at the lowest frequency which must be amplified. Therefore, a coupling capacitor usually has a high capacitance value so its capacitive reactance will be very low. Typically the coupling capacitor is an electrolytic type and has a value of 10 μF or higher.

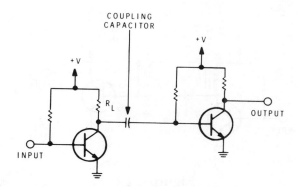

Figure 4-28
Capacitive coupling.

Capacitive-coupled amplifiers can provide amplification over a wide frequency range but there are definite upper and lower limits. The reactance of the coupling capacitor increases as frequency decreases. This means that the lower frequency limit is determined by the size of the coupling capacitor. If a higher capacitance value (thus a lower reactance) is used, it will extend the lower frequency limit. However, a lower limit of a few hertz is the best that can be obtained. The coupling capacitor will not pass DC, and it offers a large amount of reactance to extremely low frequencies.

The upper frequency limit of the capacitive-coupled stages is primarily determined by the transistor's capabilities. The gain of these devices decreases when the frequency extends beyond a certain point. Eventually, a point is reached where useful gain is no longer provided. This will be the upper frequency limit.

IMPEDANCE COUPLING

The impedance coupling technique is similar to the capacitive coupling method. However, with this method, an inductor is used in place of the the load resistor (R_L), as shown in Figure 4-29.

Impedance coupling works like capacitive coupling since the inductor (L) performs the same basic function as the load resistor, and the coupling capacitor transfers the signal from one stage to the next. However, the inductor has a very low DC resistance across its windings. The resistance of a typical inductor may be only a few hundred ohms, since the inductor is formed from low resistance wire wound around an iron core.

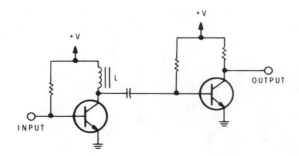

Figure 4-29
Impedance coupling.

Since the inductor has a low DC resistance, it drops only a small DC voltage. This means that the inductor itself consumes only a small amount of power. However, the inductor still offers a substantial amount of opposition to AC. Therefore, an AC signal voltage is developed across the inductor just like it would be across a load resistor; and most of this signal voltage is applied through the capacitor to the next transistor. The principle advantage that an inductor has over a resistor is that it consumes less power, thus increasing the overall efficiency of the circuit.

Unfortunately, the reactance of the inductor does not remain constant but increases with the signal frequency; and because the reactance of the inductor increases, so does the output signal voltage. This points out the main disadvantage of this type of amplifier coupling: the voltage gain gets higher as the signal frequency increases. This type of coupling is used principally in applications where only one signal frequency or a narrow range of frequencies must be amplified.

DIRECT COUPLING

Capacitive and impedance coupling techniques cannot be effectively used when very low frequencies must be amplified because the coupling capacitor cannot pass them. A technique known as "direct coupling" is used in low frequency and DC signal amplifiers.

A typical direct-coupled circuit is shown in Figure 4-30. Notice that the base of the second transistor is connected directly to the collector of the first transistor. No coupling capacitor is used between the stages.

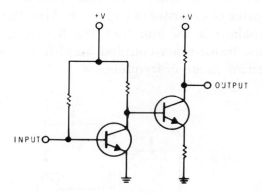

Figure 4-30
Direct coupling.

Directly coupled amplifiers provide a uniform current or voltage gain over a wide range of signal frequencies. These amplifiers may be used to amplify frequencies that range from zero (DC) to many thousands of hertz. However, they are particularly suited to low frequency applications, due to the elimination of the coupling capacitor.

Unfortunately, direct-coupled circuits are not as stable as the capacitive and impedance-coupled circuits previously described. This is because the second stage is biased by the first stage. Any changes in the output current of the first stage due to temperature variations are amplified by the second stage. This means that the operating bias of the second stage can drift extensively if the circuit is not carefully designed. To obtain a circuit that has a high degree of temperature stability, it is often necessary to use expensive, precision components or special components to compensate for temperature changes.

TRANSFORMER COUPLING

A technique known as transformer coupling is also occasionally used to couple two amplifier stages. A typical transformer coupled circuit is shown in Figure 4-31. Note that the first stage is coupled to the second stage through a transformer.

The primary advantage of transformer coupling is its ability to match differing impedances. By varying the turns ratio, the circuit designer can achieve optimum coupling between the stages. Thus, maximum power transfer takes place.

However, transformer coupling does have several disadvantages. First, the transformer is large and heavy, and therefore expensive when compared to the price of a resistor or capacitor. Also, the transformer can only pass AC signals, not DC and then the frequency range is somewhat limited. Thus, transformer-coupled amplifiers are useful only over a relatively narrow range of frequencies.

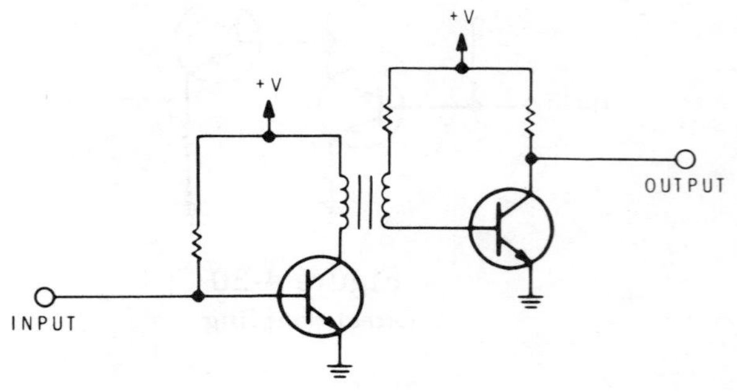

Figure 4-31
Transformer coupling.

Classes of Operation

All of the amplifier circuits we have examined so far are biased so their output currents flow during the entire cycle of the input AC voltage. As the input signal goes through its positive and negative alternations, the output current increases or decreases accordingly, but the output current always continues to flow. Any amplifier that is biased in this manner is operating as a **Class A** amplifier.

The output current of a class A amplifier is shown in Figure 4-32. Note that the current increases and decreases around its no-signal value, but never drops to zero during one complete cycle. This means that the output voltage produced by this output current also varies in a similar manner.

To operate in the class A mode, the amplifier must be biased so its output current has a no-signal value that is midway between its upper and lower limits. The upper limit is reached when the transistor is conducting as hard as it can. At this time, the output current levels off to a maximum value and the device is said to be **saturated**. Saturation occurs if the input current or voltage is increased to a sufficiently high value. Once the saturation point is reached, the output current stops increasing even though the input may continue to rise.

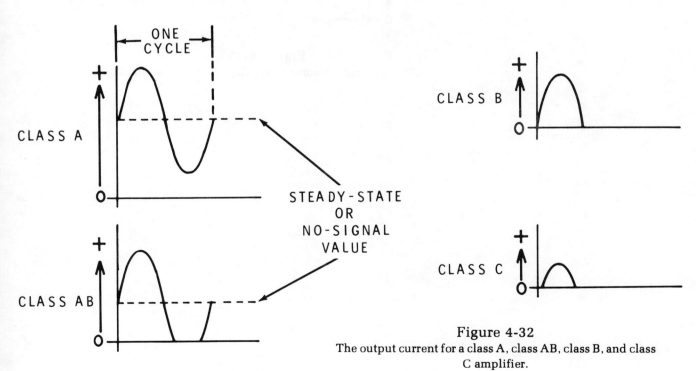

Figure 4-32
The output current for a class A, class AB, class B, and class C amplifier.

The lower limit of output current occurs when the input current or voltage is reduced so low that essentially no output current can flow. This condition is known as **cutoff**. With transistors, cutoff occurs when the input current reduces the forward bias to zero.

Therefore, class A operation is achieved by biasing the transistor midway between its saturation and cut off points. However, care must be exercised to prevent the input signal from overdriving the amplifier. If the input amplitude is so high that it causes the output current to reach the saturation and cutoff points, distortion results. If this occurs, the output signal is distorted as shown in Figure 4-33. Note that both positive and negative-going peaks are clipped off because the saturation and cutoff points are reached. However, this situation can be easily avoided if the proper input signal level is maintained.

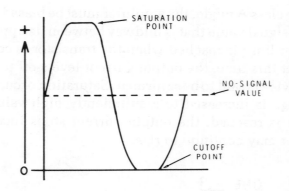

Figure 4-33
Output current distortion in a class A amplifier circuit.

When the amplifier is biased so its output current flows for less than one full cycle of the AC input signal, but for more than half of the AC cycle, the circuit is operating in the **class AB** mode. The output current waveform produced by a class AB amplifier is shown in Figure 4-32. This mode of operation occurs when the transistor is biased near the cutoff point.

You can also bias an amplifier so its output current will flow for only one half of the input AC cycle, as shown in Figure 4-32. Under these conditions, the circuit is operating in the **class B** mode. Class B operation occurs when the circuit is biased at cutoff. Under this condition, the output current can flow only on one input alteration. On the other alternation the circuit is driven further into cut off. Therefore, the class B amplifier amplifies only one half of the input AC signal.

An amplifier can also be biased so its output current flows for less than one half of the AC input cycle, as shown in Figure 4-32. Under this condition, the circuit is operating in the **class** C mode. Class C operation is obtained when the circuit is biased beyond the cutoff point. Therefore, the output current can flow only for a portion of one alternation. The voltage during this alternation must rise above cutoff before current can flow.

A class A amplifier produces an output signal that has the same basic shape and characteristics as the input signal, even though the amplitude of the output signal is higher. Class A circuits produce a minimum amount of distortion and, as a result, they are widely used in applications where a high degree of signal fidelity must be maintained.

Class AB, class B, and class C amplifiers produce a substantial amount of distortion, since these circuits amplify only a portion of the input signal. However, each of these amplifiers have specific applications in electronic equipment. They are often used in conjunction with other circuits or components that compensate for the distortion that they produce. In some applications, they are used intentionally to produce distortion in order to change the characteristics of a signal.

Amplifier Efficiency

Amplifier efficiency is a measure of how effectively an amplifier converts the DC input power (from the power supply) into AC output power. Expressed mathematically, it is:

$$\text{Efficiency in \%} = \frac{P_{out}}{P_{in}} \times 100$$

Let's try an example. What is the efficiency of a class A amplifier whose output is 10 watts while the input is 25 watts?

$$\text{Efficiency in \%} = \frac{P_{out}}{P_{in}} \times 100$$

$$\text{Efficiency in \%} = \frac{10\ W}{25\ W} \times 100$$

$$\text{Efficiency in \%} = 0.4 \times 100$$

$$\text{Efficiency} = 40\%$$

Thus, the amplifier's efficiency is 40%. That is, 40% of the DC input power is converted to AC power. Ideally, all the DC input power would be converted to AC output power for 100% efficiency.

While the class A amplifier produces the lowest distortion of any class of amplifier operation, it also has the lowest efficiency. This is due to the bias of the class A amplifier, which allows DC input current to flow whether there is an output signal or not. Therefore, the DC input power remains the same whether the amplifier produces full output or no output. Thus, the class A amplifier's efficiency will vary with the driving or input signal.

The class B amplifier's efficiency is approximately 50%. This is because it is biased at cutoff. Therefore, DC input current flows only when there is an AC input signal. Also, where there is an input signal, the amplifier is only on 50% of the time; the remainder of the time it is cut off. Thus, the class B amplifier is more efficient than the class A amplifier.

The class C amplifier has the highest efficiency of all the classes we have discussed. It is biased beyond cutoff, and the input signal must rise above cutoff to cause output current to flow. The efficiency of class C amplifiers can approach 80%.

Self-Review Questions

12. List the two types of audio amplifiers.

    1. _Voltage amplifier_
    2. _Power amplifier_

13. If the input to an amplifier is 1 mV and the output is 0.1V, what is
    the voltage gain in dB? _+40 dB_

14. If the input to an attenuator network is 10 mW and the output is 5
    mW, what is the attenuation (loss) in dB? _-3.01 dB_

15. Identify the audio amplifier coupling circuits shown in Figure
    4-34.

    A. _Capacitive Coupling_
    B. _Direct Coupling_
    C. _Transformer Coupling_
    D. _Impedance Coupling_

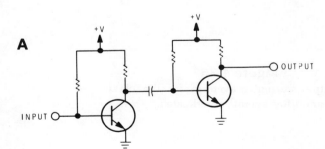

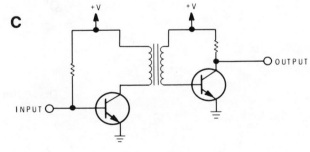

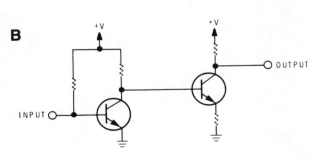

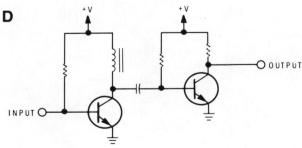

Figure 4-34
Identify these coupling circuits.

16.   Draw the output current waveform, for a sine wave input, for the classes of amplifier operation indicated in Figure 4-35.

17.   If DC input power to an amplifier is 50 watts and the AC output power is 10 watts, what is the amplifier's efficiency? _____

20%

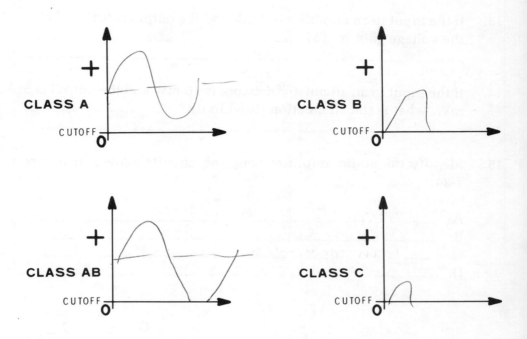

Figure 4-35
Draw the output current waveform for the classes of amplifier operation indicated.

Self-Review Answers

12.   The two types of audio amplifiers are:

  1.   Voltage amplifier.
  2.   Power amplifier.

13.   dB $= 20 \times \log \dfrac{E_2}{E_1}$

  dB $= 20 \times \log \dfrac{0.1V}{1 \text{ mV}}$

  dB $= 20 \times \log \dfrac{100 \text{ mV}}{1 \text{ mV}}$

  dB $= 20 \times \log \quad 100$

  dB $= 20 \times 2$

  dB $=$ Voltage gain is +40 dB

14.   dB $= 10 \times \log \dfrac{P_2}{P_1}$

  dB $= 10 \times \log \dfrac{10 \text{ mW}}{5 \text{ mW}}$

  dB $= 10 \times \log \quad 2$

  dB $= 10 \times 0.3010$

  dB $= 3.01$

  Attenuation is −3.01 dB

15.   The audio amplifier coupling circuits shown in Figure 4-34 are:

  A.   Capacitive coupling.
  B.   Direct coupling.
  C.   Transformer coupling.
  D.   Impedance coupling.

16.   The correct drawings are shown in Figure 4-36.

17.   Efficiency in % = $\dfrac{P_{out}}{P_{in}} \times 100$

Efficiency in % = $\dfrac{10W}{50W} \times 100$

Efficiency in % $= 0.2 \times 100$

Efficiency $=$ 20%

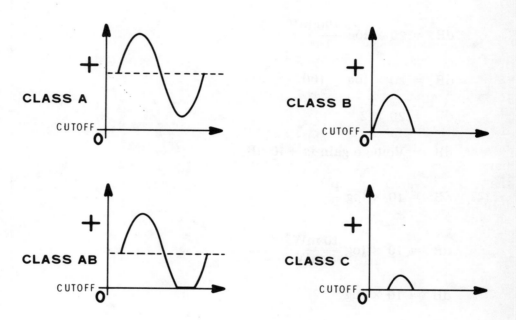

Figure 4-36
Answers to Question 16.

OPERATIONAL AMPLIFIERS

The incredibly fast development of IC technology has made the operational amplifier, or **op amp**, the most widely used and versatile of all electronic circuits. The op amp is basically an extremely high gain DC Amplifier. It was originally developed for use in "Analog" computers where it could perform a variety of functions by merely adding a few external components. In many ways, the op amp is still used in this manner. In this section, you'll discover not only what an op amp is but how it is used in several basic amplifier configurations.

The Basic Op Amp

The schematic symbol for the basic op amp is shown in Figure 4-37. Note that there are two input terminals. One is called the **inverting input** and is indicated by the minus (−) sign. The other input is called the **non-inverting input**, which is indicated by the plus (+) sign.

If a signal is applied to the inverting terminal with respect to ground, the output will be 180° out of phase with respect to the input. On the other hand, a signal applied to the non-inverting input will result in an output that is in phase with respect to the input. If signals are applied to both inputs, the output will be proportional to the difference of the two signals.

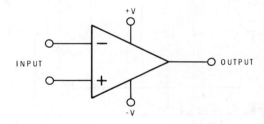

Figure 4-37
Schematic symbol for the basic op amp.

An op amp is basically a DC-coupled multistage linear amplifier. An ideal op amp would have all of the following characteristics:

- Infinite voltage gain

- Infinite input impedance

- Zero output impedance

As you might have guessed, these ideals are not reached in practice. However, op amp designs can come close to these ideals. For example, a typical op amp has a voltage gain of 200,000. Its input impedance is 2 MΩ and output impedance is 75 Ω. While these are not "ideal" characteristics, they are more than adequate for almost all applications.

An ideal op amp would also have zero output when both inputs are at zero. Any deviation from zero is called **offset**. Since a small offset usually occurs in practice, most op amps provide offset adjustment terminals. These and the other op amp terminals are shown in the IC schematic diagram of Figure 4-38. A potentiometer can be connected to the offset terminals to compensate for any offset voltage by adjusting the bias current to the input transistors.

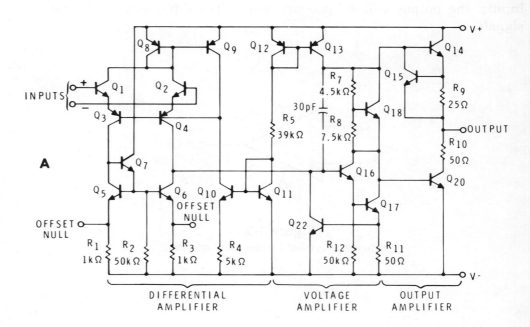

Figure 4-38
Internal circuitry of a typical op amp IC.

The Inverting Amplifier

Just as the transistor and FET have three basic circuit configurations, so too does the op amp. The inverting amplifier configuration is shown in Figure 4-39. In its most basic form, it consists of the op amp and two resistors. The noninverting input is grounded. The input signal (E_{in}) is applied through R_2 to the inverting input. The output signal is also applied back through R_1 to the inverting input. Therefore, the signal at the inverting input is determined not only by E_{in}, but also by E_{out}.

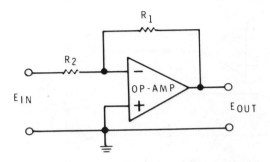

Figure 4-39
Inverting configuration. The power supply connections are not shown.

Resistor R_1 provides a **feedback** path. That is, a fraction of the output voltage is applied or "fed" back to the input. In this case, a feedback voltage is used not only to stabilize the circuit but, more importantly, to reduce the circuit voltage gain to a desired value. Since the output is 180° out of phase with the input, it subtracts from or reduces the signal present at the op amp's inverting input when it is fed back. This in turn decreases the output signal. This means that almost instantaneously, the output voltage, feedback voltage, and the voltage at the inverting input will stabilize at levels determined by the values of R_1 and R_2. Thus, by choosing the correct values of R_1 and R_2, the op amp circuit's voltage gain can be set exactly to the desired value.

The formula for the voltage gain of an inverting amplifier is:

$$A_r = \frac{R_1}{R_2}$$

Therefore, the voltage gain of the circuit shown in Figure 4-40 is:

$$A_r = \frac{R_1}{R_2} = \frac{100 \text{ k}\Omega}{10 \text{ k}\Omega}$$

$$= 10$$

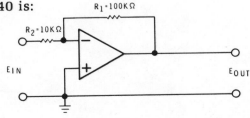

Figure 4-40
The gain of this stage is determined by R_1 and R_2.

The Noninverting Amplifier

The op amp can also be used as a noninverting amplifier. This configuration is shown in Figure 4-41. Since E_{in} is applied to the noninverting input, E_{out} is in phase with E_{in}. Therefore, in order for the feedback voltage to reduce the circuit gain, it must be applied to the inverting input. Thus, the output signal is actually the "difference" between the input signal applied to the noninverting input and the feedback signal applied to the inverting input.

Figure 4-41
The noninverting op amp configuration.

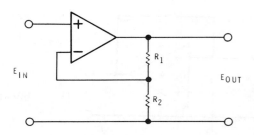

The feedback signal is obtained from the voltage divider network of R_1 and R_2. Therefore, the ratio of R_1 and R_2 determines the amount of feedback voltage and, thus, the circuit's voltage gain. In this case, the circuit's voltage gain is:

$$A_r = 1 + \frac{R_1}{R_2}$$

For example, the gain of the circuit shown in Figure 4-42 is:

$$A_r = 1 + \frac{R_1}{R_2} = 1 + \frac{100 \text{ k}\Omega}{20 \text{ k}\Omega}$$

$$= 1 + 5 = 6$$

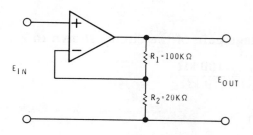

Figure 4-42
The gain of this stage is determined by R_1 and R_2.

The Voltage Follower

A special type of noninverting amplifier, a voltage follower, is shown in Figure 4-43. You will recall that the gain formula for a noninverting amplifier is:

$$A_r = 1 + \frac{R_1}{R_2}$$

In this case, $R_1 = 0$ and $R_2 = \infty$. Therefore, the voltage follower has a gain of 1. Thus, the output will be an exact replica of the input signal.

The input impedance of this circuit is extremely high, often as high as 400 MΩ, and the output impedance is very low, often less than 1Ω. Thus, the circuit behaves like a "super" emitter follower. And, like the emitter follower, it is used in impedance matching and isolation or buffering applications.

While there are many op amp circuits, almost all of them fall into the three basic circuits: inverting amplifier, noninverting amplifier, and voltage follower. This brief discussion should give you an excellent foundation for any further studies of operational amplifiers.

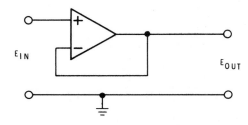

Figure 4-43
The voltage follower.

Self-Review Questions

18.   Label the terminals of the op amp shown in Figure 4-44.

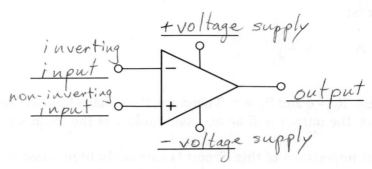

+voltage supply

inverting input

non-inverting input

output

−voltage supply

Figure 4-44
Fill in the blanks.

19.   List the characteristics of an ideal op amp. _____
infinite voltage gain
infinite input impedance
zero output impedance
zero OFFSET

20. The circuit shown in Figure 4-45A is a/an _non-inverting_ amplifier and it has a voltage gain of _11_.

21. The circuit shown in Figure 4-45B is a/an _inverting_ amplifier and it has a voltage gain of _82_.

22. The circuit shown in Figure 4-45C is a/an _voltage follower_ amplifier and it has a voltage gain of _1_.

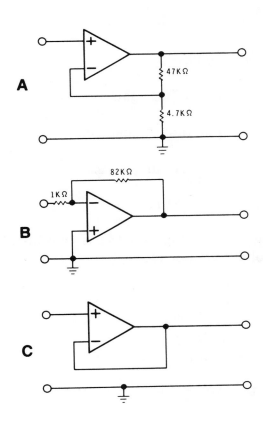

Figure 4-45
Identify these circuits and determine their gains.

Self-Review Answers

18. See Figure 4-46.

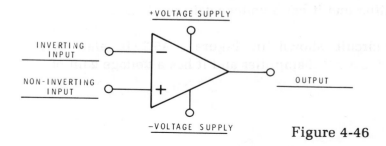

Figure 4-46

19. The ideal op amp has:

   - Infinite voltage gain.

   - Infinite input impedance.

   - Zero output impedance.

   - Zero OFFSET.

20. The circuit shown in Figure 4-45A is a **noninverting** amplifier and it has a voltage gain of **11**.

$$A_r = 1 + \frac{R_1}{R_2}$$

$$= 1 + \frac{47 \text{ k}\Omega}{4.7 \text{ k}\Omega}$$

$$= 1 + 10 = 11$$

21. The circuit shown in Figure 4-45B is an **inverting** amplifier and it has a voltage gain of **82**.

$$A_r = \frac{R_1}{R_2}$$

$$= \frac{82 \text{ k}\Omega}{1 \text{ k}\Omega} = 82$$

22. The circuit shown in Figure 4-45C is a voltage follower and it has a voltage gain of **1**.

RADIO FREQUENCY AMPLIFIERS

In this section we will continue our discussion of amplifiers. More specifically, we will be discussing radio frequency or RF amplifiers. Just as with audio amplifiers. RF amplifiers are divided into two categories: voltage amplifiers and power amplifiers. RF voltage amplifiers are used primarily in receivers, although they are also used in low level stages of transmitters. RF power amplifiers are used in transmitters.

Audio amplifiers are usually designed to amplify the whole AF spectrum, 20 to 20,000 Hz. However, radio frequency amplifiers are designed to amplify a relatively narrow portion of the RF spectrum. (The complete RF spectrum extends from 3 kHz to 300 GHz.) The portion of the spectrum that an amplifier is able to amplify is called its "bandwidth." A CB radiotelephone transmitter, for example, has a 3 kHz bandwidth, while a TV transmitter has a bandwidth of 5 to 6 MHz. Therefore, the transmitter and receiver RF amplifiers need amplify only this relatively small portion of the RF spectrum. In fact, they must reject all other frequencies.

The Tuned Amplifier

Most RF amplifiers are tuned amplifiers. They contain at least one tuned, or resonant, LC circuit. Due to the characteristics of the resonant LC circuit, the tuned amplifier amplifies only a small band of frequencies centered on the resonant frequency. The bandwidth is determined by the Q of the resonant circuit.

A tuned RF amplifier is shown in Figure 4-47. Note that it has tuned circuits on both the input and output, and that they are part of the coupling transformers. The input signal is transformer coupled to the base of the transistor Q_1. The transformer secondary (L_1) and capacitor C_1 form a parallel resonant circuit. Therefore, at the resonant frequency, maximum input voltage is developed across the base of Q_1, while little or no signal is developed at other frequencies. The output signal is developed across the tuned circuit that consists of L_2 and C_2. Again, maximum signal output is developed at the resonant frequency.

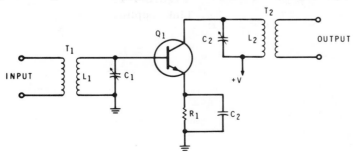

Figure 4-47
A tuned amplifier.

Coupling Circuits

Figure 4-48 shows two types of **transformer coupling**. These are also known as **inductive coupling** since they operate on the principle of mutual inductance. Figure 4-48A shows a double-tuned transformer, both primary and secondary form parallel resonant circuits. Figure 4-48B shows a single tuned transformer, only one side of the transformer is tuned.

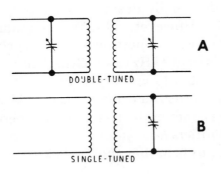

Figure 4-48
Transformer or inductive coupling.

Figure 4-49 shows a variation of transformer coupling known as **link coupling**. This type of coupling is used to connect circuits that are physically separated. As an example, link coupling could be used to couple the output of a transmitter to a separate high power amplifier.

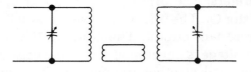

Figure 4-49
Link coupling.

Capacitive coupling is shown in Figure 4-50. The coupling capacitor in Figure 4-50A transfers RF between the two resonant circuits and their respective amplifiers. When the coupling capacitor is varied, it changes the amount of RF that is coupled from one circuit to the other.

Figure 4-50B shows a variation that is used in transistor circuits. The input impedance of the transistor is very low because it is normally forward biased. However, the impedance of the resonant LC circuit is very high. To avoid the loss of power that would come from this impedance mismatch, the coupling capacitor is connected at a lower impedance point on the inductor. This inductor "tap" is adjusted for that point where maximum power transfer occurs.

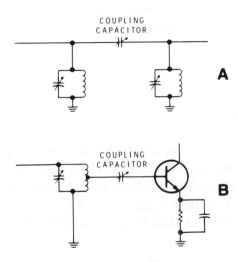

Figure 4-50
Capacitive coupling.

RF Power Amplifiers

Figure 4-51 shows a transistor RF power amplifier biased for class C operation. In this circuit, there is no base bias. However, since it takes between +0.5 and +1 volt to forward bias the emitter-base junction, the transistor is considered to be biased beyond cutoff. Therefore, it operates class C.

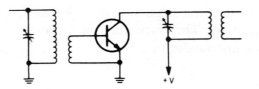

Figure 4-51
RF power amplifier.

FLYWHEEL EFFECT

Although it produces a high amount of distortion, the class C amplifier operates at a high efficiency. Therefore, RF amplifiers use an LC parallel resonant circuit, or "tank" circuit, to compensate for this distortion. The tank circuit has a "flywheel" effect which forms the complete sine wave from the pulse of current that comes from the class C amplifier.

Let's examine the flywheel effect in detail. Figure 4-52A shows a tank circuit with a battery as a current source, and a switch to control the battery current. When the switch is closed, as in Figure 4-52B, the capacitor charges with the polarity shown. Current does not flow through the inductor due to its opposition to current flow during the instant the switch is closed.

When the switch is opened, as shown in Figure 4-52C, the capacitor discharges through the inductor. This discharge current is limited by the inductor's counter emf, which is induced by the expanding magnetic field.

Once the capacitor has completely discharged, as shown in Figure 4-52D, the inductor's magnetic field starts to collapse, inducing an emf in the inductor as shown. Figure 4-52E shows that this emf and the resulting current flow recharges the capacitor in the opposite polarity. The capacitor then discharges in the opposite direction through the inductor, starting the process again. This "flywheel" action would continue indefinitely if there were no losses in the circuit.

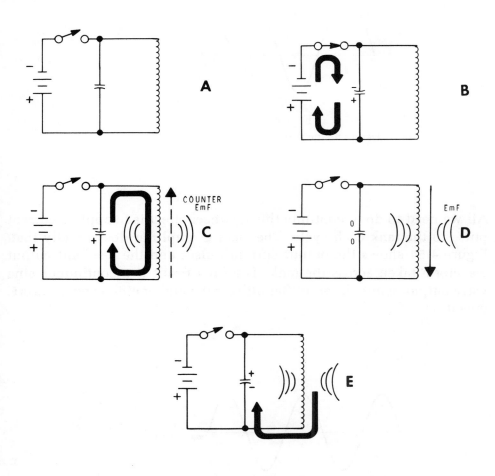

Figure 4-52
The flywheel effect.

However, due to resistance in the conductors and losses in the capacitor and inductor, the AC flywheel-effect current gradually dies out. The waveform that appears across the tank circuit is known as a damped sine wave, and is shown in Figure 4-53. The frequency of the damped sine wave is the resonant frequency of the LC tank circuit.

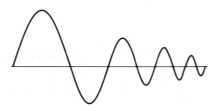

Figure 4-53
Damped sine wave for LC tank circuit.

All we need to do to continue this flywheel effect is to apply a current pulse to the tank each cycle. The class C amplifier does exactly that. Figure 4-54 shows the output current pulses and the resultant output waveform taken across the tank. The final result is a continuous sine wave output, while the amplifier utilizes the higher efficiency of class C operation.

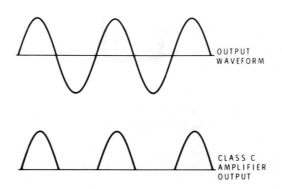

Figure 4-54
The flywheel effect completes the sine wave output signal
for the class C RF amplifier.

Tuning an RF Amplifier

The flywheel effect of the output tank circuit does indeed complete the sine wave; but the tank circuit must also be tuned to exactly the right frequency for maximum output. A typical RF amplifier is shown in Figure 4-55. C_1 and L_1 form the resonant tank circuit, and M_1 is a meter used to indicate the collector current of Q_1.

When a signal is applied to the input, Q_1 amplifies it and the output voltage is developed across the impedance of C_1 and L_1 in parallel. When C_1 and L_1 are in resonance, their impedance is maximum. Therefore maximum voltage is developed across the tank and coupled to the output.

To tune this amplifier, you would adjust C_1, so that the tank circuit resonates at the input signal's frequency. You can do this by monitoring the collector current of Q_1, or the output voltage, while adjusting C_1. When the tank circuit is adjusted to resonance at the input frequency, the output voltage will be maximum and the collector current will be minimum (due to the high impedance of the tank circuit at resonance).

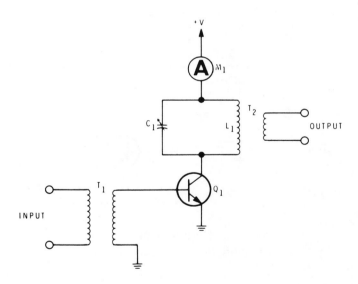

Figure 4-55
RF amplifier.

Self-Review Questions

23.   Identify the RF amplifier coupling circuits shown in Figure 4-56.

A.   _Link coupling_
B.   _Transformer or inductive coupling_
C.   _Capacitive coupling_

24.   An RF amplifier can be operated class C due to the _____ _Flywheel effect_ _____ of the LC tank circuit.

25.   When you tune an RF amplifier, resonance is indicated by maximum _output voltage_ and minimum _collector current_

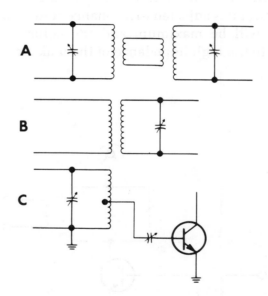

Figure 4-56
Identify these coupling circuits.

Self-Review Answers

23. The RF amplifier coupling circuits shown in Figure 4-56 are:

    A. Link coupling.
    B. Transformer or inductive coupling.
    C. Capacitive coupling.

24. An RF amplifier can be operated class C due to the **flywheel effect** of the LC tank circuit.

25. When you tune an RF amplifier, resonance is indicated by maximum **output voltage** and minimum **collector or plate current.**

OSCILLATORS

An oscillator is a circuit that generates a repetitive AC signal. The frequency of this AC signal may be as low as a few hertz or as high as several gigahertz.

There are many applications for oscillators. They are an extremely important part of every transmitter and receiver. In fact, each modern day transmitter or receiver has at least two oscillators and sometimes more. A TV receiver has at least four oscillators. Oscillators are also required for RF signal generators, audio signal generators, and many other test instruments.

There are almost as many different types of oscillators as there are applications for them. In this section, we will be discussing several of the most commonly used oscillators. However, all oscillators operate on the same basic principles. If you thoroughly understand these basic principles, you should be able to analyze the operation of almost all oscillators.

Oscillator Requirements

Basically, an oscillator is an amplifier that gets its input signal from its own output. That is, a portion of the output signal is fed back to the input to maintain oscillation. This feedback signal must be **regenerative**, or in phase, with the input signal. Compare this to the op amp whose feedback voltage is 180° out of phase. This is called **degenerative** feedback, since it reduces the amplifier gain. On the other hand, regenerative feedback increases amplifier gain and it sustains oscillation.

Figure 4-57 shows a basic block diagram of a common emitter amplifier with regenerative feedback. Since the amplifier has a 180° phase shift the feedback network must also introduce a 180° phase shift so that the feedback signal will be in phase with the input.

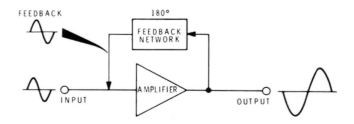

Figure 4-57
Common emitter amplifier with regenerative feedback.

Once the amplifier begins to operate, you can remove the input signal and the circuit will continue to oscillate. The gain of the amplifier replaces energy lost in the circuit, and regenerative feedback sustains oscillation.

However, this circuit could oscillate on any number of frequencies. Even minor noise pulses could change the frequency of this kind of oscillator. Therefore, it is necessary to have a means of setting the oscillator frequency.

To do this, the frequency selectivity of the parallel LC network is useful because it resonates at a specific frequency, determined by the values of inductance and capacitance. Also, the required 180° phase shift is produced across this network. Therefore, an LC tank circuit in the feedback loop, as shown in Figure 4-58, will control the frequency. The tank resonates at its natural frequency and amplifier gain replaces energy lost in the tank.

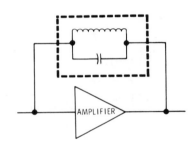

Figure 4-58
An LC network controls the frequency.

Up to this point, the oscillators were amplifiers with input signals applied to start the circuit. In actual practice, oscillators must start on their own. This is a natural phenomenon. When a circuit is first turned on, energy levels do not instantly reach maximum, but gradually approach it. This produces noise pulses that are phase shifted and fed back to the input, as shown in Figure 4-59. The amplifier steps up these pulses, which are again supplied to the input. This action continues and oscillation is underway. Therefore, the oscillator is naturally self-starting.

Let's summarize the requirements for an oscillator:

1. An amplifier is necessary to replace circuit losses.

2. Frequency determining components are needed to set the oscillation frequency.

3. Regenerative feedback is required to sustain oscillations.

4. The oscillator must be self-starting.

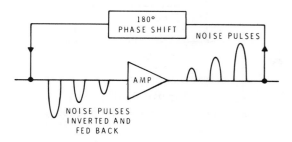

Figure 4-59
Oscillators are self-starting.

The Armstrong Oscillator

The "Armstrong," or "tickler coil" feedback, oscillator is shown in Figure 4-60. This circuit uses mutual inductance or transformer action between L_1 and tickler coil L_2 for regenerative feedback. L_1 and C_1 determine the frequency of operation, and transistor Q1 is the amplifier.

When voltage is applied to the circuit, current flows through Q1 and the tickler coil L_2. This current produces an expanding magnetic field around L_2. This, in turn, induces voltage into L_1, which starts "flywheel effect" oscillations in the tank circuit of L_1-C_1. These oscillations are applied to the base of Q_1 and amplified. The amplified signal is developed across L_2 and, through transformer action, fed back to the resonant tank circuit L_1-C_2; thus sustaining the oscillations.

To maintain the oscillation, the amplifier must supply enough gain to overcome the circuit losses. You can vary the amount of feedback voltage by adjusting the position of the tickler coil with respect to the tank coil. Note that since the transistor introduces a 180° phase shift, the transformer must also supply a 180° phase shift for regenerative feedback to occur.

Notice that the base connection for Q_1 is "tapped" down on L_1. This is done to obtain a low impedance to match the transitor's low base-emitter impedance and therefore obtain maximum power transfer.

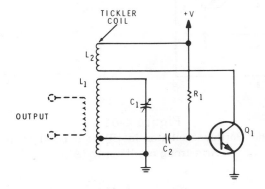

Figure 4-60
Transistor Armstrong oscillator.

Hartley Oscillator

The Hartley oscillator is actually an adaptation of the basic Armstrong oscillator shown in Figure 4-61A. Here, L_2 is the tickler coil. C_2 couples the feedback signal to the tickler coil and prevents the DC voltage from being shorted to ground.

One of the undesirable features of the Armstrong oscillator is that the tickler coil has a tendency to resonate with the distributed capacitance in the circuit. This results in oscillator frequency variations. Figure 4-61B shows you that, by making the tickler coil part of the resonant tank, we can eliminate this undesirable effect. Now, both L_1 and L_2 resonate with C_1. This circuit is known as a Hartley oscillator and you can easily recognize it by the tapped coil in the tank circuit.

The Hartley oscillator operates in a manner that is similar to the Armstrong oscillator. Regenerative feedback is still accomplished by "tickler coil" L_2, which is now part of the resonant tank.

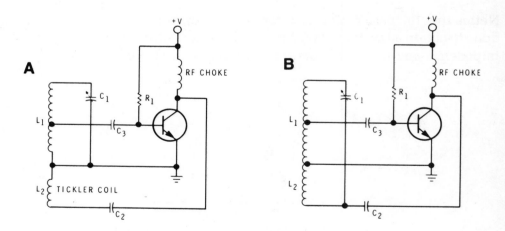

Figure 4-61
How the Hartley oscillator in (B) is derived from the
Armstrong oscillator in (A.)

Figure 4-62 shows two types of transistor Hartley oscillators. You can see that both are Hartley oscillators, by the tapped coil, but the DC current path differs between the two. Figure 4-62A is called a "series-fed Hartley oscillator" because the path for DC current is from ground through L_2, which is part of the tank, and through Q_1 to the power supply.

Figure 4-62B is called a "shunt-fed Hartley oscillator." Its DC current path is from ground through Q_1 to the power supply. Therefore, a series fed oscillator's DC current path is through part of the tank and the transistor, while a shunt-fed oscillator's DC current flows only through the transistor. The disadvantage of the series-fed oscillator is that the DC current heats L_2 causing it to expand and change value. This, in turn, causes the oscillator frequency to change. Therefore, the shunt-fed Hartley has better frequency stability.

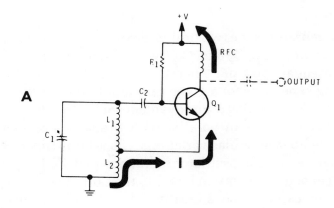

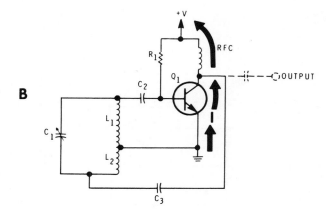

Figure 4-62
(A) Series-fed Hartley oscillator,
(B) shunt-fed Hartley oscillator.

Colpitts Oscillator

You can easily identify the Colpitts oscillator, shown in Figure 4-63, by the two capacitors, C_1 and C_2, in the resonant tank. The operation is very similar to the Hartley, with the exception that the capacitive branch of the tank is tapped rather than the inductive branch. Regenerate feedback voltage is developed across C_2, which sustains the flywheel oscillations of the tank circuit. Therefore, the Colpitts oscillator utilizes capacitive feedback.

The resonant frequency of the Colpitts oscillator, or any LC oscillator, is determined by the usual formula for resonance:

$$f_r = \frac{1}{2\pi \sqrt{LC}}$$

However, C represents the two tuning capacitors, C_1 and C_2, connected in series. For example, if C_1 and C_2 are each 400 pF, then the combined capacitance equals 200 pF. Therefore, the Colpitts oscillator requires larger tuning capacitors for a given resonant frequency than other oscillator circuits. This is actually an advantage. Since the transistor's internal junction capacitance is in parallel with the tank, any changes in the transistor, due to heat or bias variations, causes this capacitance to vary. This, in turn, causes the oscillator frequency to drift. However, because the tuning capacitors are much larger in the Colpitts oscillator, the effect of the junction capacitance is greatly reduced. This is a decided advantage for the Colpitts oscillator.

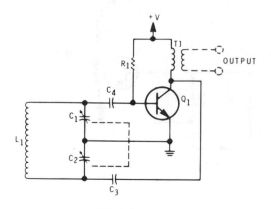

Figure 4-63
Transistor Colpitts oscillator.

Crystal Oscillators

Certain crystalline substances, such as quartz, have an unusual electrical characteristic. If mechanical pressure is applied to the crystal, a DC voltage is generated. Likewise, if a vibrating mechanical pressure is applied, an AC voltage is generated. Conversely, if an AC voltage is applied across the crystal, the crystal undergoes a physical change, resulting in mechanical vibration. This relationship between electrical and mechanical effects is known as the **piezoelectric effect**.

In its natural shape, quartz is a hexagonal prism with pyramids at the ends. Slabs are cut from the "raw" quartz to obtain a usable crystal. There are many ways to cut a crystal, all with different names; such as X cut, NT cut, CT cut, and AT cut. Each cut has a different piezoelectric property. Figure 4-64 shows how the most popular cut, the AT, is obtained from the quartz crystal.

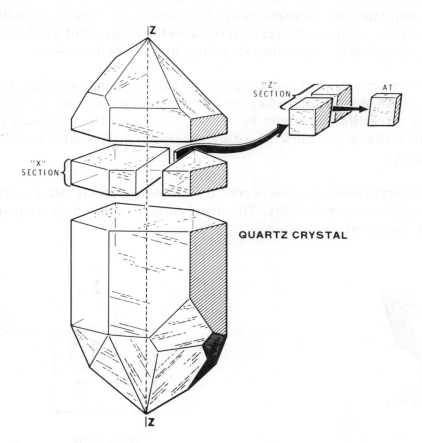

QUARTZ CRYSTAL

Figure 4-64
How an "AT" cut is obtained from a quartz crystal.

Because of their structure, crystals have a natural frequency of vibration. If the frequency of the applied AC signal matches this natural frequency, the crystal will vibrate a large amount. However, if the frequency of the exciting voltage is slightly different than the crystal's natural frequency, little vibration is produced. The crystal, therefore, is extremely frequency selective, making it desirable for filter circuits. Also, the crystal's mechanical frequency of vibration is extremely constant, which makes it ideal for oscillator circuits.

The natural frequency of a crystal is usually determined by its thickness. As shown in Figure 4-65, the thinner the crystal is, the higher is its natural frequency. Conversely, the thicker a crystal is, the lower is its natural frequency. To obtain a specific frequency, the crystal slab is ground to the required dimensions. Of course there are practical limits on just how thin a crystal can be cut, without it becoming extremely fragile.

Crystals are used in a number of oscillator circuits. Figure 4-66 shows a crystal Colpitts oscillator, which is sometimes called a Pierce oscillator. C_3 provides the path for regenerative feedback, while C_1 and C_2 identify it as a Colpitts oscillator. The crystal replaces the tank circuit's inductor. It also determines the operating frequency and holds it constant.

The major advantage of using crystals is their frequency stability. However, unlike the LC oscillators we have previously discussed, the frequency of oscillation cannot be varied. To change the frequency of a crystal oscillator, you must use a different crystal that is cut to the desired frequency.

In summary: The advantages of crystal oscillators are their extreme precision and frequency stability. The disadvantage is that their frequency cannot be varied.

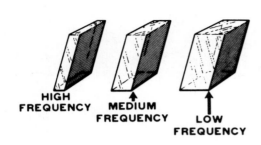

Figure 4-65
Thickness determines the crystal's natural frequency.

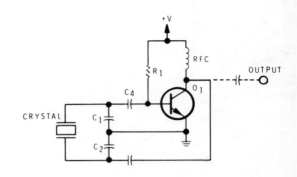

Figure 4-66
Crystal Colpitts oscillator.

Self-Review Questions

26. What are the requirements for an oscillator? _____

_____

1. must have amplifier
2. must have frequency
   determining components
3. must have regenerative
   feed back
4 must be self-starting

27. Identify the schematic diagrams shown in Figure 4-67.

   A. _____*Armstrong oscillator*_____
   B. _____*Colpitts oscillator*_____
   C. _____*Crystal colpitts oscillator*_____
   D. _____*Hartley oscillator*_____

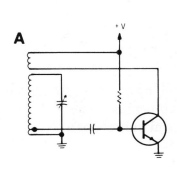
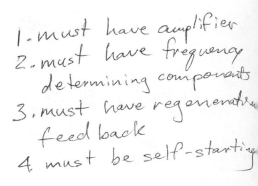
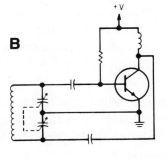

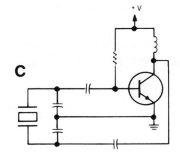
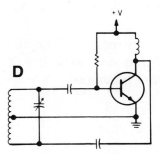

Figure 4-67
Identify these oscillators.

28. What is the "piezoelectric effect"? _____*voltage generated*_
   *when pressure applied to crystal*_

29. What determines the natural frequency of a crystal? _____*thickness*_
   _____*of crystal*_____

30. What are the advantages of using a crystal oscillator when compared to an LC oscillator? What are the disadvantages? _____

   _____

1. precise frequency
   selection
2. stable frequency

can't vary
frequency.

Self-Review Answers

26. The requirements for an oscillator are as follows:

    1. It must have amplifier to replace circuit losses.
    2. It must have frequency determining components to set the oscillation frequency.
    3. It must have regenerative feedback to sustain oscillations.
    4. It must be self-starting.

27. The schematic diagrams shown in Figure 4-67 are:

    A. Armstrong oscillator.
    B. Colpitts oscillator.
    C. Crystal Colpitts oscillator.
    D. Hartley oscillator.

28. The piezoelectric effect is exhibited by quartz crystals. When pressure is applied to the crystal, a voltage is generated. Conversely, when a voltage is applied, the crystal undergoes a physical change, resulting in mechanical vibration.

29. The thickness of a crystal determines its natural frequency. The thinner the crystal is, the higher is its natural frequency.

30. The advantages of crystal oscillators are their extreme precision and frequency stability. The disadvantage is that their frequency cannot be varied.

APPENDIX A

Logarithms

Since pocket calculators and home computers have become an everyday part of our lives, the use of logarithms to simplify problem solving has diminished. However, to properly understand decibels, you need some knowledge of logarithms.

What is a logarithm? In the very simplest terms, it is an exponent. The **common** logarithm is an exponent of the base 10. It is not necessary that 10 be used for a base. Any number could be used; for instance, the system of **natural** logarithms uses 2.71828 for a base. Two could be used, or eight, or any other number. In this discussion, we will limit our study to the **common** logarithms which use the base 10. In your study of electronics, you have already done considerable work with "powers of 10."

When we express a number in powers of 10, its exponent is the logarithm of the number. For example:

$$1 = 10^0, \text{ or } 0 \text{ is the logarithm of } 1.$$
$$10 = 10^1, \text{ or } 1 \text{ is the logarithm of } 10.$$
$$100 = 10^2, \text{ or } 2 \text{ is the logarithm of } 100.$$
$$1000 = 10^3, \text{ or } 3 \text{ is the logarithm of } 1000.$$

Characteristic and Mantissa

Every number has a logarithm. Numbers such as 100 or 1,000 have whole number logarithms since they are exact multiples of the base 10. However, the number 32, for instance, is between 10 and 100. Its logarithm must be between 1 (the log of 10) and 2 (the log of 100). Actually, the log of 32 is 1.5.

With logarithms the whole number is called the **characteristic**, and the fractional part is the **mantissa**. In our example of 1.5 the logarithm consists of a characteristic of 1, and a mantissa of 0.5.

Finding a Log of a Number

The first step in finding the log of a number is to determine the characteristic. How the number fits in the scale of 1, 10, 100, 1000, and so on determines its characteristic. For example, the log of 1 is zero, thus any number less than 1 will have a negative characteristic.

Since the log of 1 is zero and the log of 10 is 1, the log of any number between 1 and 10 will have a characteristic of zero. For instance, 5 has a log of 0.6990. The log of 100 is 2; therefore, the log of any number between 10 and 100 has a characteristic of 1. For instance, 50 has a log of 1.6990. The log of a number from 100 up to 1000 will have a characteristic of 2. For instance, the log of 500 is 2.6990.

In logarithms of 5, 50, and 500, only the characteristic changed. The mantissa, the part of the logarithm following the decimal point, is the same for each number. The mantissa is the same for any given sequence of digits regardless of decimal point placement.

How do we find the mantissa? The easiest way to find both characteristic and mantissa is with a scientific calculator. However, if you do not have such a calculator, you can look up the mantissa in Table 1. Here is the way to find a logarithm of a number using the table.

Find the logarithm of 354.

Since 354 is between 100 and 1000, you know that the characteristic is 2. Turn to Table 1, and go down the extreme left hand column of numbers until you locate 35. Next, move across the page until you intersect the column marked 4. Column numbers are at both the top and bottom of the page. The mantissa of 354 is 5490. Combining this with the characteristic gives 2.5490 which is the logarithm of 354.

Try another. What is the logarithm of 6?

When you look in the column of numbers you see that there is no number less than 10. However, this doesn't matter. Remember that the mantissa of 5 was the same as the mantissa of 50 or 500. Thus the mantissa of 6 is the same as the mantissa of 600. So, go down the numbers column to 60 then across the 0 column and read the mantissa of 7782. Since 6 is between 0 and 10 the characteristic is 0; therefore, the logarithm of 6 is 0.7782.

You have seen that any number may be expressed as a power of 10 and that the exponent involved is called the common logarithm of that number. The logarithm has two parts. The characteristic precedes the decimal point and is used to indicate the location of the decimal point in the original number. The characteristic is determined by inspection. The second part of the logarithm, the mantissa, follows the decimal point and indicates the numerical value of the original number. The mantissa is found from a table.

TABLE 1

LOGARITHMS OF NUMBERS

	0	1	2	3	4	5	6	7	8	9
10	0000	0043	0086	0128	0170	0212	0253	0294	0334	0374
11	0414	0453	0492	0531	0569	0607	0645	0682	0719	0755
12	0792	0828	0864	0899	0934	0969	1004	1038	1072	1106
13	1139	1173	1206	1239	1271	1303	1335	1367	1399	1430
14	1461	1492	1523	1553	1584	1614	1644	1673	1703	1732
15	1761	1790	1818	1847	1895	1903	1931	1959	1987	2014
16	2041	2068	2095	2122	2148	2175	2201	2227	2253	2279
17	2304	2330	2355	2380	2405	2430	2455	2480	2504	2529
18	2553	2577	2601	2625	2648	2672	2695	2718	2742	2765
19	2788	2810	2833	2856	2878	2900	2923	2945	2967	2989
20	3010	3032	3054	3075	3096	3118	3139	3160	3181	3201
21	3222	3243	3263	3284	3304	3324	3345	3365	3385	3404
22	3424	3444	3464	3483	3502	3522	3541	3560	3579	3598
23	3617	3636	3655	3674	3692	3711	3729	3747	3766	3784
24	3802	3820	3838	3856	3874	3892	3909	3927	3945	3962
25	3979	3997	4014	4031	4048	4065	4082	4099	4116	4133
26	4150	4166	4183	4200	4216	4232	4249	4265	4281	4298
27	4314	4330	4346	4362	4378	4393	4409	4425	4440	4456
28	4472	4487	4502	4518	4533	4548	4564	4579	4594	4609
29	4624	4639	4654	4669	4683	4698	4713	4728	4742	4757
30	4771	4786	4800	4814	4829	4843	4857	4871	4886	4900
31	4914	4928	4942	4955	4969	4983	4997	5011	5024	5038
32	5051	5065	5079	5092	5105	5119	5132	5145	5159	5172
33	5185	5198	5211	5224	5237	5250	5263	5276	5289	5302
34	5315	5328	5340	5353	5366	5378	5391	5403	5416	5428
35	5441	5453	5465	5478	5490	5502	5514	5527	5539	5551
36	5563	5575	5587	5599	5611	5623	5635	5647	5658	5670
37	5682	5694	5705	5717	5729	5740	5752	5763	5775	5786
38	5798	5809	5821	5832	5843	5855	5866	5877	5888	5899
39	5911	5922	5933	5944	5955	5966	5977	5988	5999	6010
40	6021	6031	6042	6053	6064	6075	6085	6096	6107	6117
41	6128	6138	6149	6160	6170	6180	6191	6201	6212	6222
42	6232	6243	6253	6263	6274	6284	6294	6304	6314	6325
43	6335	6345	6355	6365	6375	6385	6395	6405	6415	6425
44	6435	6444	6454	6464	6474	6484	6493	6503	6513	6522
45	6532	6542	6551	6561	6571	6580	6590	6599	6609	6618
46	6628	6637	6646	6656	6665	6675	6684	6693	6702	6712
47	6721	6730	6739	6749	6758	6767	6776	6785	6794	6803
48	6812	6821	6830	6839	6848	6857	6866	6875	6884	6893
49	6902	6911	6920	6928	6937	6946	6955	6964	6972	6981
50	6990	6998	7007	7016	7024	7033	7042	7050	7059	7067
51	7076	7084	7093	7101	7110	7118	7126	7135	7143	7152
52	7160	7168	7177	7185	7193	7202	7210	7218	7226	7235
53	7243	7251	7259	7267	7275	7284	7292	7300	7308	7316
54	7324	7332	7340	7348	7356	7364	7372	7380	7388	7396
	0	1	2	3	4	5	6	7	8	9

Courtesy of U.S. Coast and Geodetic Survey

Table 1 (continued)

LOGARITHMS OF NUMBERS.

	0	1	2	3	4	5	6	7	8	9
55	7404	7412	7419	7427	7435	7443	7451	7459	7466	7474
56	7482	7490	7497	7505	7513	7520	7528	7536	7543	7551
57	7559	7566	7574	7582	7589	7597	7604	7612	7619	7627
58	7634	7642	7649	7657	7664	7672	7679	7686	7694	7701
59	7709	7716	7723	7731	7738	7745	7752	7760	7767	7774
60	7782	7789	7796	7803	7810	7818	7825	7832	7839	7846
61	7853	7860	7868	7875	7882	7889	7896	7903	7910	7917
62	7924	7931	7938	7945	7952	7959	7966	7973	7980	7987
63	7993	8000	8007	8014	8021	8028	8035	8041	8048	8055
64	8062	8069	8075	8082	8089	8096	8102	8109	8116	8122
65	8129	8136	8142	8149	8156	8162	8169	8176	8182	8189
66	8195	8202	8209	8215	8222	8228	8235	8241	8248	8254
67	8261	8267	8274	8280	8287	8293	8209	8306	8312	8319
68	8325	8331	8338	8344	8351	8357	8363	8370	8376	8382
69	8388	8395	8401	8407	8414	8420	8426	8432	843	8445
70	8451	8457	8463	8470	8476	8482	8488	8494	8500	8506
71	8513	8519	8525	8531	8537	8543	8549	8555	8561	8567
72	8573	8579	8585	8591	8597	8603	8609	8615	8621	8627
73	8633	8639	8645	8651	8657	8663	8669	8675	8681	8686
74	8692	8698	8704	8710	8716	8722	8727	8733	8739	8745
75	8751	8756	8762	8768	8774	8779	8785	8791	8797	8802
76	8808	8814	8820	8825	8831	8837	8842	8848	8854	8859
77	8865	8871	8876	8882	8887	8893	8899	8904	8910	8915
78	8921	8927	8932	8938	8943	8949	8954	8960	8965	8971
79	8976	8982	8987	8993	8998	9004	9009	9015	9020	9025
80	9031	9036	9042	9047	9053	9058	9063	9069	9074	9079
81	9085	9090	9096	9101	9106	9112	9117	9122	9128	9133
82	9138	9143	9149	9454	9159	9165	9170	9175	9180	9186
83	9191	9196	9201	9206	9212	9217	9222	9227	9232	9238
84	9243	9248	9253	9258	9263	9269	9274	9279	9284	9289
85	9294	9299	9304	9309	9315	9320	9325	9330	9335	9340
86	9345	9350	9355	9360	9365	9370	9375	9380	9385	9390
87	9395	9400	9405	9410	9415	9420	9425	9430	9435	9440
88	9445	9450	9455	9460	9465	9469	9474	9479	9484	9489
89	9494	9499	9504	9509	9513	9518	9523	9528	9533	9538
90	9542	9547	9552	9557	9562	9566	9571	9576	9581	9586
91	9590	9595	9600	9605	9609	9614	9619	9624	9628	9633
92	9638	9643	9647	9652	9657	9661	9666	9671	9675	9680
93	9685	9689	9694	9699	9703	9708	9713	9717	9722	9727
94	9731	9736	9741	9745	9750	9754	9759	9763	9768	9773
95	9777	9782	9786	9791	9795	9800	9805	9809	9814	9818
96	9823	9827	9832	9836	9841	9845	9850	9854	9859	9863
97	9868	9872	9877	9881	9886	9890	9894	9899	9903	9908
98	9912	9917	9921	9926	9930	9934	9939	9943	9948	9952
99	9956	9961	9965	9969	9974	9978	9983	9987	9991	9996
	0	1	2	3	4	5	6	7	8	9

Unit 5

DIGITAL ELECTRONICS

CONTENTS

INTRODUCTION

The purpose of this unit on digital electronics is to give you an overview of the subject and to introduce you to its basic concepts. You will learn the basic digital logic functions; elementary Boolean Algebra; and apply these techniques in several experiments. However; this unit is merely an extended introduction to digital electronics. To fully understand digital techniques; we highly recommend Heath Individual Learning Program EE-3201, Digital Techniques.

Examine the Unit Objectives listed in the next section to see what you will learn in this unit.

UNIT OBJECTIVES

When you have completed this unit you should be able to:

1. State why the binary number system is used in digital electronics.

2. Convert between the decimal, binary, and binary-coded-decimal (BCD) number systems.

3. Identify each of the six basic logic gate symbols, their truth tables, and their logic expressions.

4. Write the truth table and logic expressions for each of the six basic logic elements.

5. Define the term "flip-flop" and name three basic types.

6. Identify the logic diagrams and symbols of the three basic types of flip-flops.

7. Answer questions about and discuss the basic operation of the three types of flip-flops.

8. Name the two most widely used types of sequential logic circuits.

9. Identify binary up, binary down, and BCD counters from their logic diagrams.

10. Answer questions about and discuss the basic operation of counters and shift registers.

11. Define the terms "monostable" and "astable."

12. Determine the output pulse duration of a 555 monostable multivibrator.

13. Determine the output pulse duration of a 555 astable multivibrator.

NUMBER SYSTEMS

Many different number systems are being used throughout the world each day. But the first system that you always think of is the decimal system, the one we all learned in school and still use daily. It has ten digits — 0 through 9.

You also use the "duodecimal" system each day, which uses the symbols (or digits) 1 through 12. Since the day is divided into two 12-hour periods, and the year is divided into 12-month periods, you use this system to record the passage of time.

There are also many other number systems in use. In this section, you will learn about the ones that are used in digital electronics and computers. But first, we'll review the basic principles of the decimal system.

The Decimal System

Most of us are very familiar with the decimal number system, where the digits 0 through 9 are combined in a certain way to indicate a specific quantity. For example, the number 1984 is made up of 4 digits. The right-hand digit (the least significant digit, or LSD) stands for a number times ten raised to the power of zero, or 10^0. You will recall that 10^0, or any number raised to the power of zero, is equal to 1 (See "Appendix A," "Scientific Notation," in Unit 1). The next digit represents a number times 10^1, and this continues until the left-hand, (the most significant digit or MSD) is reached. Therefore, you can express this number as:

$$1984 = (1 \times 10^3) + (9 \times 10^2) + (8 \times 10^1) + (4 \times 10^0) =$$
$$1000 + 900 + 80 + 4$$

Thus, the decimal number system is known as a positional, or weighted, system. Each digit position in a number carries a particular weight in determining the magnitude of that number.

If you tried to design an electronic circuit that would respond to these decimal numbers, you would need a different voltage level to represent each one of the ten digits. The circuit would have to be capable of both generating and detecting each of these ten levels. This would be difficult to do and would require very complex circuitry.

However, consider a system where you could represent all of the numbers you need (only two in this case) by only the absence or the presence of a single voltage; usually, a simple "on or off" state. This can be easily accomplished with a much smaller number of electronic components: a simple switch is either on or off; a transistor is either conducting or cut off; an LED is either on or off. This number system, which has only two characters or digits, is called the "binary number system."

The Binary Number System

The two digits that are used in the binary system are 0 and 1. And since there are only two digits, the binary system is based on powers of two. For example, the binary number 1101 can be broken down in the following manner:

$$1101 = \quad (1 \times 2^3) \quad + \quad (1 \times 2^2) \quad + \quad (0 \times 2^1) \quad + \quad (1 \times 2^0)$$

This number can then be written in decimal notation as:

$$8 \quad + \quad 4 \quad + \quad 0 \quad + \quad 1 \quad = 13$$

In a binary number, each digit is called a **bit**. This is actually a contraction of the words binary digit.

As you can see, it takes a lot more binary digits or "bits" to represent a given quantity than it does to represent the same quantity using the decimal system. However, it is much more convenient to use the binary system in electronic circuits, since there are a large number of components that have two distinct levels of operation.

Number System Conversions

Since electronic circuits use the binary number system exclusively, you must be capable of converting binary numbers to decimal and vice versa.

To convert a binary number into its decimal equivalent, you add together the weights of the positions in the number where binary 1's occur. The weights of the number positions are shown below.

2^7	2^6	2^5	2^4	2^3	2^2	2^1	2^0
128	64	32	16	8	4	2	1

As an example, convert the binary number 1010 into its decimal equivalent. You can add the weights of the positions where 1's occur as shown below.

Binary Number	1		0		1		0		
Position Weights	(8)		(4)		(2)		(1)		
Decimal Equivalent	8	+	0	+	2	+	0	=	10

Next, convert the binary number 101101 to its decimal equivalent.

Binary Number	1		0		1		1		0		1		
Position Weights	(32)		(16)		(8)		(4)		(2)		(1)		
Decimal Equivalent	32	+	0	+	8	+	4	+	0	+	1	=	45

After you solve a few practice problems you will quickly catch on to this procedure.

To convert a decimal number into binary, you must repeatedly divide the number by 2 and note the remainder. The remainder, which will always be either 1 or 0, forms the equivalent binary number.

As an example, convert the number 175 into its binary equivalent.

<u>Remainder</u>

$175 \div 2 = 87$	1 ◄——— Least significant Bit
$87 \div 2 = 43$	1
$43 \div 2 = 21$	1
$21 \div 2 = 10$	1
$10 \div 2 = 5$	0
$5 \div 2 = 2$	1
$2 \div 2 = 1$	0
$1 \div 2 = 0$	1 ◄——— Most significant Bit

Therefore, the decimal number 175 is equal to 10101111 in binary. You can check it by converting the binary version back into decimal form using the procedure we discussed earlier.

Binary Coded Decimal

Even though we know what the binary number system is, it is still difficult to deal with. This is simply because we are unfamiliar with it. And, let's face it; its cumbersome when you have to deal with large numbers. However, the binary system is the natural language of the two-state digital electronic components. Therefore, to simplify the reading and handling of large numbers, special binary codes have been developed.

These special codes convert each decimal digit to a four-bit binary number. This can be easily accomplished because it takes four binary digits to represent the decimal numbers 0 through 9. Therefore, a two-digit decimal number will require 8 bits, a three-digit decimal number — 12 bits, and so on. This code is known as the **binary coded decimal** (BCD) system.

The most commonly used BCD system uses the first ten binary numbers (0-9) and rejects the remaining six (10-15) that the four bits are capable of representing. Therefore, it has ten valid codes and six invalid codes. This code is called the natural BCD code or the 8421 code.

To represent a decimal number in BCD notation, you substitute the appropriate four-bit code for each decimal digit. For example, the number 937 in BCD would be:

Decimal Number —	9	3	7
Binary Code —	1001	0011	0111

The BCD number would be 1001 0011 0111. A space is left between each four-bit group so there will be no danger of confusion between the BCD code and the "pure" binary code.

The beauty of the BCD code is that the ten BCD code combinations are very easily remembered. Once you begin to work with binary numbers on a regular basis, you will find that the BCD numbers will come to you as quickly and automatically as decimal numbers. For that reason by simply glancing at the BCD representation of a decimal number you can make the conversion almost as quickly as if it were already in decimal form.

Self-Review Questions

1. What is the base of the binary number system? _____.

2. Why is the binary number system used in digital electronics? ___
   _____
   _____

3. Convert the following binary numbers to decimal.

   A. 11001
   B. 1110010
   C. 1110101
   D. 1101101

4. Convert the following decimal numbers to binary.

   A. 89
   B. 52
   C. 101
   D. 156

5. Convert the following BCD numbers to decimal.

   A. 1001 0011
   B. 1000 0101
   C. 0110 0101 0010
   D. 0100 0000 1001

6. Convert the following decimal numbers to BCD.

   A. 42
   B. 69
   C. 705
   D. 923

Self-Review Answers

1. The base of the binary number system is 2.

2. The binary number system is used in digital electronics because a large number of electronic devices have two distinct operating states, such as "on and off", etc.

3. A — Binary Number   1   1   0   0   1

    Position Weights (16)   (8)   (4)   (2)   (1)

    Decimal Number   16 + 8 + 0 + 0 + 1 = 25

    B — 114
    C — 117
    D — 109

4. A — 89 ÷ 2 = 44   Remainder   1 (LSB)
        44 ÷ 2 = 22   0
        22 ÷ 2 = 11   0
        11 ÷ 2 = 5   1
          5 ÷ 2 = 2   1
          2 ÷ 2 = 1   0
          1 ÷ 2 = 0   1 (MSB)

    Binary Number = 1011001

    B — 110100
    C — 1100101
    D — 10011100

5. A — BCD Number   1001   0011
         Decimal Number   9   3   = 93

    B — 85
    C — 652
    D — 409

6. A — Decimal Number   4   2
         BCD   0100   0010
                  BCD = 0100 0010

    B — 0110 1001
    C — 0111 0000 0101
    D — 1001 0010 0011

LOGIC ELEMENTS

All digital equipment, simple or complex, is constructed from just a few basic circuits. These circuits are called logic elements. A logic element performs some specific logic function on binary data.

There are two basic types of digital logic circuits: decision making and memory. Decision making logic elements monitor binary inputs and produce outputs based on the input states and on the characteristics of the logic element itself. Memory logic elements are used to store binary data. This section discusses the basic decision making logic elements.

The Inverter

The simplest form of logic element is the inverter, or NOT circuit. The inverter is a logic element whose output state is always the opposite of its input state. If the input is a binary 0, the output is a binary 1. If the input is a binary 1, the output is a binary 0. The inverter has an output that is the **complement** of the input. The binary states 1 and 0 are considered to be complementary.

The operation of the inverter, or any logic element, is summarized by a chart known as a **truth table**. This chart shows all of the possible input states and the resulting outputs.

Figure 5-1 is the truth table for an inverter. The input is designated A, while the output is labeled $\overline{A}$, which is pronounced as "A NOT" or "A BAR". The bar over the letter A indicates the complement of A. Note that the truth table shows all possible input combinations and the corresponding output for each. Since the inverter has a single input, there are only two possible input combinations: 0 and 1. The output in each case is the complement or opposite of the input.

INPUT	OUTPUT
A	$\overline{A}$
0	1
1	0

Figure 5-1
Truth table for logic inverter.

The symbols used to represent a logic inverter are shown in Figure 5-2. The triangle portion of the symbol represents the circuit itself, while the circle designates the inversion or complementary nature of the circuit. Either of the two symbols may be used. Note the input and output labeling. Such simplified symbols are used instead of the actual electronic schematic in order to simplify the drawing and application of a logic circuit. It is the logic function and not the circuit that is important. For this reason, we will not be examining the actual circuitry of logic elements. Exactly how the logic operation is performed is not important. It is more important to know what each logic element is, what it does, and how to use it.

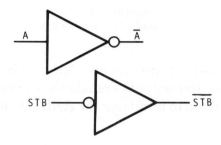

Figure 5-2
Symbols for a logic inverter.

The AND Gate

The AND gate is a logic circuit that has two or more inputs and a single output. The output of this gate is a binary 1 if, and only if, all inputs are binary 1. If any one or more inputs are a binary 0, the output will be binary 0. For example, in a two-input gate, a one input AND a one input results in a one output; hence the name AND gate.

The operation of a two-input AND gate is summarized in the truth table of Figure 5-3. The inputs are designated A and B. The output is designated C. The output for all possible input combinations is indicated in the truth table.

INPUTS		OUTPUT
A	B	C
0	0	0
0	1	0
1	0	0
1	1	1

Figure 5-3
Truth table for an AND gate.

The basic symbol used to represent an AND gate is shown in Figure 5-4. Note that the inputs and outputs are labeled to correspond to the truth table in Figure 5-3. Keep in mind that the AND gate may have any number of logical inputs.

An important point to note about the AND symbol in Figure 5-4 is the equation at the output, $C = A \cdot B$ or $C = AB$. This equation is a form of algebraic expression that is used to designate the logical function being performed. The equation expresses the output C in terms of the input variables A and B. It is read as, "C equals A AND B." Here the AND function is designated by the dot between the two input variables A and B.

Figure 5-4
Logic symbol for an AND gate.

The OR Gate

Another basic logic element is the OR gate. Like the AND gate, it can have two or more inputs and a single output. Its operation is such that the output is a binary 1 if any one or all inputs are a binary 1. The output is a binary 0 only when both inputs are binary 0.

The logical operation of a two input OR gate is expressed by the truth table in Figure 5-5. The truth table designates all four possible input combinations and the corresponding output. Note that the output is a binary 1 when either or both of the inputs are binary 1. That is, the output is 1 if input D OR input E OR both are present.

INPUTS		OUTPUT
D	E	F
0	0	0
0	1	1
1	0	1
1	1	1

Figure 5-5
Truth table for an OR gate.

The logic symbol for an OR gate is shown in Figure 5-6. The inputs are labeled according to the truth table in Figure 5-5. Notice that the output expression for the OR gate is $F = D + E$. The plus sign is used to designate the logical OR function.

Figure 5-6
Logic symbol for an OR gate.

The NAND Gate

The term NAND is a contraction of the expression NOT-AND. Therefore, a NAND gate is an AND gate followed by an inverter. Figure 5-7A shows the basic diagram of a NAND gate. Note the algebraic output expression for the AND gate and the inverter. The entire AND output expression is inverted and indicated by the bar over it.

Figure 5-7B shows the standard symbol used for a NAND gate. It is similar to the AND symbol but a circle has been added to the output to represent the inversion that takes place.

The logical operation of the NAND gate is easy to deduce from the circuit in Figure 5-7. Its operation is indicated by the truth table in Figure 5-8. The NAND output is simply the complement of the AND output.

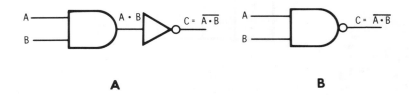

A   **B**

Figure 5-7
A NAND gate.

INPUTS		OUTPUT	
A	B	AND $A \cdot B$	NAND $\overline{A \cdot B} = C$
0	0	0	1
0	1	0	1
1	0	0	1
1	1	1	0

Figure 5-8
Truth table for a NAND gate.

The NOR Gate

Like the term NAND, NOR is a contraction for the expression NOT-OR. Therefore, the NOR gate is essentially a circuit that combines the logic functions of an OR gate and an inverter.

Figure 5-9A is a logic representation of a NOR gate. Figure 5-9B shows the standard symbol used to represent a NOR gate. Note that the output expression is the inverted OR function.

The logical operation of the NOR gate is illustrated by the truth table of Figure 5-10. The NOR output is simply the complement of the OR function. Like any other logic gate, both NAND and NOR gates may have two or more inputs, as required by the application.

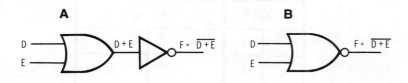

Figure 5-9
A NOR gate.

INPUTS		OUTPUT	
D	E	OR D+E	NOR $\overline{D+E}$=F
0	0	0	1
0	1	1	0
1	0	1	0
1	1	1	0

Figure 5-10
Truth table for a NOR gate.

The Exclusive OR Gate

The standard OR gate is generally referred to as an inclusive OR. The OR circuit produces a binary 1 output if any one or more of its inputs are binary 1. However, the exclusive OR produces a binary 1 output only if one, and only one, of the inputs is a binary 1. The truth table shown in Figure 5-11 compares the output of the standard inclusive OR and the exclusive OR.

A special symbol is used to designate the exclusive OR function in logic expressions. Just as the plus sign represents OR and the dot represents the AND function, the symbol $\oplus$ represents the exclusive OR function. The exclusive OR of inputs A and B is expressed as:

$$C = A \oplus B$$

The symbol for an exclusive OR gate is shown in Figure 5-12.

INPUTS		OUTPUTS	
A	B	STANDARD OR	EXCLUSIVE OR
0	0	0	0
0	1	1	1
1	0	1	1
1	1	1	0

Figure 5-11
Truth table for both inclusive and exclusive OR gates.

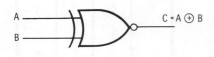

Figure 5-12
The Exclusive OR symbol.

Self-Review Questions

7. Identify the logic symbols shown in Figure 5-13.

A. _____   D. _____
B. _____   E. _____
C. _____   F. _____

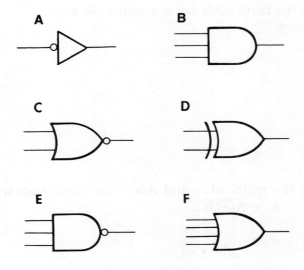

Figure 5-13
Identify these symbols.

8. If the input to a logic inverter is labeled RSA, the output will be _____.

9. Write the logic expression for the circuit of Figure 5-14.

_____.

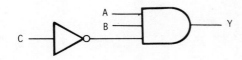

Figure 5-14
What is the logic expression for this circuit?

10. Write the truth table for a 3-input AND gate.

11. Write the truth table for a 4-input OR gate.

12. Write the truth table and draw the logic diagram for the expression $Y = \overline{A \oplus B}$.

13. Which of the following logic expressions matches the circuit shown in Figure 5-15?

A.	X =	$\overline{E + \overline{F} + \overline{G}}$
B.	X =	$E + F + G$
C.	X =	$\overline{E \cdot \overline{F} \cdot \overline{G}}$
D.	X =	$\overline{\overline{E} \cdot F \cdot G}$

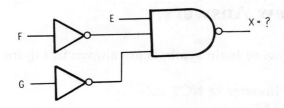

Figure 5-15
What is the logic expression for this circuit?

14. Which of the following truth table outputs indicates the operation of the circuits shown in Figure 5-15.

INPUTS			OUTPUT X = ?			
E	F	G	A	B	C	D
0	0	0	1	1	0	0
0	0	1	1	1	0	1
0	1	0	1	1	0	1
0	1	1	1	1	0	1
1	0	0	1	0	1	1
1	0	1	1	1	0	1
1	1	0	1	1	0	1
1	1	1	0	1	0	1

15. What logic function is indicated by the truth table shown below?

INPUT		OUTPUT
A	B	C
0	0	1
0	1	0
1	0	0
1	1	0

Self-Review Answers

7. The following logic symbols are shown in Figure 5-13:

        A. Inverter or NOT gate.
        B. AND gate.
        C. NOR gate.
        D. Exclusive OR gate.
        E. NAND gate.
        F. OR gate.

8. If the input to a logic inverter is labeled RSA, the output will be $\overline{RSA}$.

9. The logic expression for the circuit of Figure 5-14 is:

$$Y = A \cdot B \cdot \overline{C}$$

10. The truth table for a 3-input AND gate is:

INPUTS			OUTPUT
A	B	C	D
0	0	0	0
0	0	1	0
0	1	0	0
0	1	1	0
1	0	0	0
1	0	1	0
1	1	0	0
1	1	1	1

11. The truth table for a 4-input OR gate is:

INPUTS				OUTPUT
A	B	C	D	E
0	0	0	0	0
0	0	0	1	1
0	0	1	0	1
0	0	1	1	1
0	1	0	0	1
0	1	0	1	1
0	1	1	0	1
0	1	1	1	1
1	0	0	0	1
1	0	1	0	1
1	0	1	1	1
1	1	0	0	1
1	1	0	1	1
1	1	1	0	1
1	1	1	1	1

12. The logic diagram and truth table for $Y = \overline{A \oplus B}$ is shown in Figure 5-16.

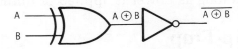

INPUTS		OUTPUT
A	B	Y
0	0	1
0	1	0
1	0	0
1	1	1

Figure 5-16

13. C — The logic expression for Figure 5-15 is:

$$X = \overline{E \cdot \overline{F} \cdot \overline{G}}$$

14. B — Output B indicates the operation of the circuit shown in Figure 5-15.

15. The truth table indicates the NOR logic function.

FLIP-FLOPS

The output signals of the logic elements we just discussed depend only on the instantaneous value of the input signals. These elements are known as "combinatorial logic elements." Their operation does not depend on the history of previous states, because they have no memory function.

Elements whose operation depend on their history are called "sequential elements." The most important element of this class is the flip-flop. In this section, we will discuss the types of flip-flops that are used most often in sequential logic and digital memory circuits.

What is a Flip-Flop?

A flip-flop is a digital logic circuit whose basic function is memory, or storage. A flip-flop is capable of storing a single bit of binary data. It can assume either of two stable states, one representing a binary 1 and the other a binary 0. If the flip-flop is put into one of its two stable states, it will remain there as long as power is applied or until it is changed.

The Latch Flip-Flop

There are three basic types of flip-flops: the latch, the D type, and the JK. Let's start with the simplest, the latch or set-reset flip-flop, the simplest form of binary storage element. The symbol shown in Figure 5-17 is used to represent this type of flip-flop.

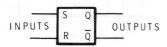

Figure 5-17
The latch flip-flop.

The flip-flop has two inputs, S and R, and two outputs Q and $\overline{Q}$. Applying the appropriate logic signal to either the S or R input will put the latch into one state or the other. The S input is used to **set** the flip-flop. When a flip-flop is set, it is said to be storing a binary 1. The R input is used to **reset** the flip-flop. A reset flip-flop is said to be storing a binary 0.

The latch has two outputs, labeled Q and $\overline{Q}$. These are called the **normal** and **complement** outputs respectively. As in other logic circuits, you can use any letter or alphanumeric combination to designate logic symbols. For example, the designation FF2, meaning flip-flop number 2, could be used as shown in Figure 5-18.

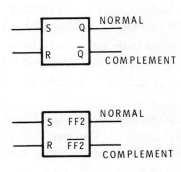

Figure 5-18
Alphanumerics can be used to label the normal and complement outputs.

To tell what state the flip-flop is in, you look at the normal output. The logic level there tells you which bit, 0 or 1, is being stored. At the same time, the complement output has the state opposite that of the normal output.

A latch can easily be constructed with two NAND gates, as shown in Figure 5-19. Here, the two gates are wired back-to-back so that the output of one feeds the input to the other.

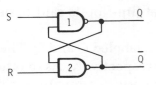

Figure 5-19
A NAND gate latch.

If the S and R inputs are both binary 1, the normal condition for the latch, the circuit is storing a bit put there earlier. For example, if the flip-flop is set, the normal (or Q) output from gate 1 will be at binary 1. This output is fed back to the upper input of gate 2. The lower input to gate 2 is also a binary 1, so its output ($\overline{Q}$) is low or binary 0.

The output from gate 2 is fed back to the lower input of gate 1. This input holds the Q output high (binary 1). You can see why this circuit is called a latch. Because of the feedback arrangement, the flip-flop is latched into this state. It will stay this way until you change it. The way you change it is by applying a low level to either the set or reset inputs.

If a binary zero or low level is applied to the R input of the latch, it will "flip" to the reset condition. The low level on the R input will force the output of gate 2 high. This will cause both inputs to gate 1 to be high, so its output will be low or binary 0. This indicates that the flip-flop is in the reset state. Any further low levels applied to the reset input will have no effect.

If a low level is applied to the S input, the latch will "flop" to the set condition. You can verify this by following the logic levels through the circuit of Figure 5-19. Therefore, the latch is like a toggle switch, it is either in one position or the other. And, once the change-over is made, repeating the action has no further effect.

However, with both S and R inputs low, the Q and $\overline{Q}$ outputs will **both** be high. No longer are the outputs complementary. Therefore, we really don't know what state the latch is in. It is in an ambiguous state and is neither set nor reset. This condition is one of the pecularities of a latch. When you are using it, you have to be careful to avoid simultaneous low inputs on the S and R terminals.

The operation of a latch can be summarized by the truth table of Figure 5-20, which accounts for all possible input and output states. Note that when both S and R inputs are binary 1, the output state of the flip-flop is designated X, where X can be either a 1 or a 0 as determined by previous input conditions. This is known as the "store" condition.

INPUTS		OUTPUTS		
S	R	Q	$\overline{Q}$	STATE
0	1	1	0	SET
1	0	0	1	RESET
1	1	X	$\overline{X}$	SET OR RESET "STORE" CONDITION
0	0	1	1	AMBIGUOUS CONDITION

Figure 5-20
A latch truth table.

D Type Flip-Flop

The symbol for the D type flip-flop is shown in Figure 5-21. Like any other flip-flop, it has two outputs that are used to determine its contents, that is, the outputs indicate what bit is stored there. For example, if the $\overline{Q}$ output is high, the Q or normal output is the complement or a binary 0. The Q output tells you the state of the flip-flop directly. Since it is a binary 0 in this example, the flip-flop is reset. The point here is that you read the outputs of a D flip-flop just as you would a latch or, for that matter, any other flip-flop.

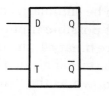

Figure 5-21
The D type flip-flop.

Now look at the inputs. Just as on the latch, there are two. However, these work differently. The D input is where you apply the data or bit to be stored. Of course, it can be either a binary 1 or a binary 0. The T input line controls the flip-flop. It is used to determine whether the input data is recognized or ignored. If the T input line is high (binary 1), the data on the D input line is stored in the flip-flop. If the T line is low (binary 0), the D input line is not recognized. The bit stored previously is retained. The D line can essentially do anything and it will be ignored if T is low.

You can get a better idea of how the D flip-flop works by taking a look at its circuitry. The logic diagram of one type of D flip-flop is shown in Figure 5-22.

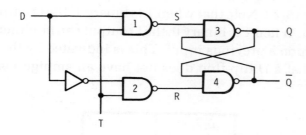

Figure 5-22
A D type flip-flop logic diagram.

In this circuit, gates 3 and 4 form a latch where the bit is stored. Gates 1 and 2 are enabling gates that pass or inhibit the input. The inverter makes sure that the S and R inputs to the latch are always complementary to avoid any possibility of the ambiguous state occurring.

Suppose a binary 1 is applied to the D input. If the T input is low (binary 0), the outputs of NAND gates 1 and 2 will be held at binary 1 regardless of the D input. This is the normal state for the inputs of a latch. Therefore, the latch will be undisturbed. Thus, a low on the T input prevents the flip-flop's state from changing and effectively "disconnects" the D input.

When the T input goes high, the D input determines the outputs of gates 1 and 2 and, therefore, the state of the latch. If the D input is high, the inputs to gate 1 will both be high and the output will go low. The top input to gate 2 will be low due to the inverter. Therefore, the S input of the latch will be low and R input, high. The latch will go to the set condition. Thus, a binary 1 on the D input, along with a binary 1 on the T input, will set the flip-flop. If the T input now goes low, the D input will be disabled and the latch will remain in the set condition.

What happens when T input is high and the D input is low? In this case, both inputs to gate 2 will be high and its output will go low. The output of gate 1 will remain high. Therefore, the latch will reset.

The operation of a D type flip-flop is completely described by the truth table of Figure 5-23. Note that when T is binary 1, the Q output is the same as the D input. When T is binary 0, the Q output can be either binary 0 or 1, depending upon a previous input. This is indicated by the X state in the table. Note that a D flip-flop does not have an ambiguous state.

INPUTS		OUTPUTS	
D	T	Q	$\bar{Q}$
0	0	X	$\bar{X}$
0	1	0	1
1	0	X	$\bar{X}$
1	1	1	0

Figure 5-23
A D flip-flop truth table.

JK Flip-Flops

The JK flip-flop is the most versatile type of binary storage element in common use. It can perform all of the functions of the RS and D type flip-flops, plus several other things. Naturally, it is more complex and expensive than the other types, so for that reason it isn't always used where simpler and less expensive circuits will do.

A JK flip-flop is really two flip-flops in one. It usually consists of two latches, one feeding the other, with appropriate input gating on each. This is shown in Figure 5-24.

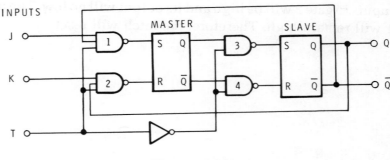

Figure 5-24
A JK flip-flop.

This arrangement is called a master-slave JK flip-flop. The master flip-flop is the input circuit. Logic signals applied to the JK flip-flop set or reset this master latch. However, the slave flip-flop is the latch from which the outputs are taken. The slave latch gets its input from the master latch. Both latches are controlled by a clock pulse*. Since there are **two** places to store bits in a JK flip-flop, there can be times when both master and slave latches are identical or times when they are complementary. But only the slave latch is responsible for indicating the state of the JK flip-flop.

The input of the master latch is controlled by gates 1 and 2. Gates 3 and 4 control the transfer of the master latch state to the slave latch. Input T, which is the clock signal, controls the input gating circuits just as in the D flip-flop. The inverter keeps the clock to the master and slave input gates complementary. The clock pulse controls the JK flip-flop while the J and K inputs determine exactly how it will be controlled. You will also see the T input referred to as CP or CK, designating clock pulse or clock.

Now let's see exactly how the J, K, and T (clock) inputs affect the flip-flop. Consider the time when the clock input is low. Gates 1 and 2 will be "inhibited" and their outputs will remain high. Therefore, the J and K inputs cannot control the state of the master latch when the T input is low.

The slave latch will have the same state as the master latch when the clock input is low. This is because the output of the inverter in the clock line is binary 1, causing gates 3 and 4 to be enabled during this time. Therefore, the state of the master latch is transferred to the slave latch. For example, if binary 1 is stored in the master latch, both inputs to gate 3 will be high, resulting in a low output. Gate 4 will have a low input from the master latch and, thus, a high output. You will recall that a low on the S input and a high on the R input of a latch causes it to set or to store a binary 1. Thus, the binary 1 has been transferred from the master latch to the slave latch.

Now if the clock T goes high, gates 1 and 2 will be enabled. The output of the inverter will inhibit gates 3 and 4. Therefore, the master latch cannot further change the slave latch. However, with gates 1 and 2 now enabled, the J and K inputs can affect the state of the master latch.

*A periodic signal that causes logic circuits to be stepped from one state to the next. These signals will be described in detail later in "Clocks and One Shots".

If both J and K inputs are low, the outputs of gates 1 and 2 will be held high, so no change takes place in the master latch.

If the J and K inputs are both high, then the state of the master latch will be determined by the Q and $\overline{Q}$ outputs, which are fed back to gates 1 and 2. For example, if the slave latch is set, the master latch will then be reset. If the slave is reset, the master will be set. The reason for this is the way the outputs are criss-crossed back to gates 1 and 2. Thus, with the J, K, and T inputs high, the state of the master latch will be determined by the Q and $\overline{Q}$ outputs.

The J and K inputs are similar to the set and reset inputs of a latch. If J is 1 and K is 0, the master latch will be set. If J is 0 and K is 1, the master latch will be reset. Remember, the T input line must be high for this to happen. When the T input goes low, gates 1 and 2 will be inhibited, while gates 3 and 4 will be enabled. Therefore, the contents of the master latch will be transferred to the slave latch the instant the T input goes low.

The operation of a JK flip-flop is summed up in the truth table of Figure 5-25. Note that only the normal (Q) output condition is shown. However, it is given twice, once prior to a clock pulse (t) and then after one clock pulse (t + 1). Output state X can represent either set (1) or reset (0); in any case, it represents the state that was previously stored in the flip-flop.

INPUTS		OUTPUTS	
J	K	Q (t)	Q (t + 1)
0	0	X	X
0	1	X	0
1	0	X	1
1	1	X	$\overline{X}$

Figure 5-25
A JK flip-flop truth table.

Now, consider each of its input conditions. First, if both J and K inputs are at binary 0, input gates 1 and 2 will be disabled at all times and the JK flip-flop will retain its previous condition, an inhibit or store mode. If K is high (1) and J is low (0), when the clock (T) pulse goes high the master latch will be reset. When the clock pulse goes low, this reset condition will be transferred to the slave latch. Therefore, the flip-flop will be reset and the normal output (Q) will be a binary 0.

If J is high (1) and K is low (0), when the clock pulse goes high the master latch will be set. When the clock pulse goes low, the set condition will be transferred to the slave latch. The flip-flop is now set and the Q output will be a binary 1.

If both J and K are high, the feedback signals from the Q and $\overline{Q}$ outputs will affect the flip-flop. For example, if the flip-flop is set, a clock pulse will cause the master latch to reset. This occurs because the Q and $\overline{Q}$ signals are criss-crossed (see Figure 5-24). When the clock pulse goes low, the reset condition will be transferred to the slave latch. Now, if the J and K inputs remain high, the flip-flop will continue to complement itself with each clock pulse. This is known as the "toggle" mode of JK flip-flop operation.

The waveforms for the toggle mode are shown in Figure 5-26. Notice that the flip-flop only changes state when the T input goes from high to low; this corresponds to the transfer of data from the master to the slave latch. Also note that the output frequency is one-half the input frequency. Therefore, in the toggle mode, the JK flip-flop can be used as a two-to-one frequency divider. Several JK flip-flops can be cascaded to permit division by any factor of 2 (2, 4, 8, 16, 32, etc.)

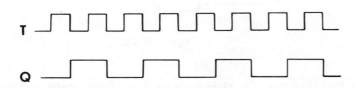

Figure 5-26
Waveforms for a JK flip-flop in the toggle mode.

This completes the description of basic, JK flip-flop operation. The symbol used to represent it is shown in Figure 5-27.

Figure 5-27
The JK flip-flop symbol.

Self-Review Questions

16. What is a flip-flop? _____

_____

17. Name the three types of flip-flops.

1. _____
2. _____
3. _____

18. What type of flip-flop is shown in Figure 5-28?
_____.

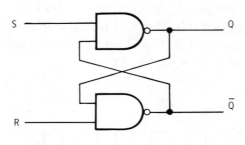

Figure 5-28

19. In Figure 5-28, if both inputs are low, what state will the flip-flop be in? What logic levels will appear at the Q and $\overline{Q}$ outputs?

_____

_____

20. What is the normal output of a flip-flop if it is set?

_____

21. The complement output of a flip-flop is low. What is the bit value stored? _____

22. What type of flip-flop is shown in Figure 5-29?
_____

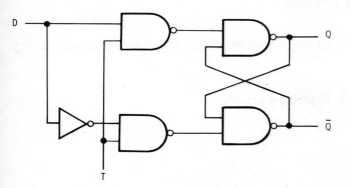

Figure 5-29

INPUTS		OUTPUTS	
D	T	Q	$\bar{Q}$
0	0		
0	1		
1	0		
1	1		

Figure 5-30

23. In Figure 5-29, if the T input is high and the D input is low, what will be the state of the complement output? What will be the state of the normal output?

_____

_____

_____

24. Complete the truth table shown in Figure 5-30 for the D type flip-flop.

25. On a JK flip-flop, the J input is high and the K input is low. What is the state of the flip-flop after one clock pulse occurs on the T input?

_____

26. The state of the JK flip-flop changes when the clock signal on the T input switches from _____.
   high to low/low to high

27. If both J and K inputs are low and the Q output is 0, what is the state of the flip-flop after three clock pulses? _____

28. If both J and K inputs are high and the $\bar{Q}$ output is 0, what is the state of the flip-flop after five clock pulses? _____

_____

29. In a JK flip-flop, both J and K inputs are at binary 1. The T input is a 50 kHz square wave. What is the Q output? _____

Self-Review Answers

16. A flip-flop is a digital logic circuit whose basic function is memory or storage.

17. The three types of flip-flops are the:

    1. Latch.
    2. D type.
    3. JK.

18. A NAND gate latch is shown in Figure 5-28.

19. If both inputs to a NAND latch are low, it will be in an ambiguous output state. Both the Q and $\overline{Q}$ outputs will be high.

20. Normal output of a set flip-flop is a binary 1.

21. If the complement output is low, the normal output is high. Therefore, a binary 1 is stored.

22. A D type flip-flop is shown in Figure 5-29.

23. If the T input is high and the D input is low, the complement output will be high and the normal output will be low.

24. See Figure 5-31.

INPUTS		OUTPUTS	
D	T	Q	$\overline{Q}$
0	0	X	$\overline{X}$
0	1	0	1
1	0	X	$\overline{X}$
1	1	1	0

Figure 5-31

25. With the J input high and K input low, after one clock pulse the flip-flop will be set.

26. The state of the JK flip-flop changes when the clock signal on the T input switches from **high to low**.

27. If both the J and K inputs are low, the flip-flop will be in the inhibit or store mode. Therefore, the flip-flop will not change state regardless of the number of clock pulses. The Q output will remain at 0.

28. If both the J and K inputs are high, the flip-flop will toggle. The $\overline{Q}$ output after five clock pulses will, therefore, be a binary 1.

29. With both the J and K inputs high, the flip-flop will toggle and divide the clock frequency by 2. The Q output will be a 25 kHz square wave.

COUNTERS AND SHIFT REGISTERS

A binary counter is a sequential logic circuit made up of flip-flops, that is used to count the number of binary pulses applied to it. The pulses or logic level transitions to be counted are applied to the counter input. These pulses cause the flip-flops in the counter to change state in such a way that the binary number stored in the flip-flops is representative of the number of input pulses that have occurred. By observing the flip-flop outputs, you can determine how many pulses were applied to the input.

There are several different types of counters used in digital circuits. The most commonly used is the binary counter. This circuit counts in the standard pure binary code. BCD counters, which count in the standard 8421 BCD code, are also widely used. In addition, both up and down counters are available.

The shift register is another widely used type of sequential logic circuit. Like a counter, it is made up of binary storage elements, usually flip-flops. These storage elements are cascaded in such a way that the bits stored there can be moved or shifted from one element to another adjacent element.

All of the storage registers in a shift register are actuated simultaneously by a single input clock, or shift, pulse. When a shift pulse is applied, the data stored in the shift register is moved one position in one of two directions. The shift register is basically a storage medium, where one or more binary words may be stored. However, its ability to move the data one bit at a time from one storage element to another makes the shift register valuable in performing a wide variety of logic operations. The following pages will give you detailed descriptions of shift registers and binary counters.

Binary Counters

A binary counter is a sequential logic circuit that uses the standard pure binary code. Such a counter is made up by cascading JK flip-flops, as shown in Figure 5-32. The normal output of one flip-flop is connected to the toggle (T) input of the next flip-flop. The JK inputs on each flip-flop are open or high. The input pulses to be counted are applied to the toggle input of the A flip-flop.

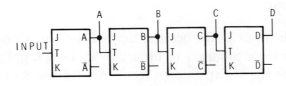

Figure 5-32
A 4-bit binary counter.

To see how this binary counter operates, remember that a JK flip-flop toggles, or changes state, each time a trailing edge transition occurs on its T input. The flip-flops will change state when the normal output of the previous flip-flop switches from binary 1 to binary 0. If we assume that the counter is initially reset, the normal outputs of all the flip-flops will be binary 0. When the first input pulse occurs, the A flip-flop will become set. The binary number stored in the flip-flops indicates the number of input pulses that have occurred. To read the number stored in the counter, you simply observe the normal outputs of the flip-flops. The A flip-flop is the least significant bit of the word. Therefore, the four-bit number stored in the counter is designated DCBA. After this first input pulse, the counter state is 0001. This indicates that one input pulse has occurred.

When the second input pulse occurs, the A flip-flop toggles and, this time, becomes reset. As it resets, its normal output switches from binary 1 to binary 0. This causes the B flip-flop to become set. Observing the new output state, you see that it is 0010, or the binary equivalent of the decimal number 2. Two input pulses have occurred.

When the third input pulse occurs the A flip-flop will again set. The normal output switches from binary 0 to binary 1. This transition is ignored by the T input of the B flip-flop. The number stored in the counter at this time then is 0011, or the number 3, indicating that three input pulses have occurred.

When the fourth input pulse occurs, the A flip-flop is reset. Its normal output switches from binary 1 to binary 0, thereby toggling the B flip-flop. This causes the B flip-flop to reset. As it does, its normal output switches from binary 1 to binary 0, causing the C flip-flop to become set. The number now in the counter is 0100, or a decimal 4. This process continues as the input pulses occur. The count sequence is the standard 4-bit binary code, as indicated in Figure 5-33.

D	C	B	A	
0	0	0	0	
0	0	0	1	
0	0	1	0	
0	0	1	1	
0	1	0	0	
0	1	0	1	
0	1	1	0	
0	1	1	1	RECYCLE
1	0	0	0	
1	0	0	1	
1	0	1	0	
1	0	1	1	
1	1	0	0	
1	1	0	1	
1	1	1	0	
1	1	1	1	

Figure 5-33
Count sequence for a 4-bit binary counter.

An important point to consider is the action of the circuit when the number stored in the counter is 1111. This is the maximum value of a four bit number and the maximum count capacity of the circuit. When the next input pulse is applied, all flip-flops will change state. As the A flip-flop resets, the B flip-flop resets. As the B flip-flop resets it, in turn, resets the C flip-flop. As the C flip-flop resets, it toggles the D flip-flop, which is also reset to zero. The result is that the contents of the counter becomes 0000. As you can see from Figure 5-33, when the maximum content of the counter is reached, it simply recycles and starts its count again.

The complete operation of the four-bit binary counter is illustrated by the input and output waveforms in Figure 5-34. The upper waveform is a series of input pulses to be counted. Here they are shown as a periodic binary waveform; but, of course, it is not necessary for the input signal to be of a constant frequency or have equally spaced input pulses. The waveforms also show the normal output of each flip-flop.

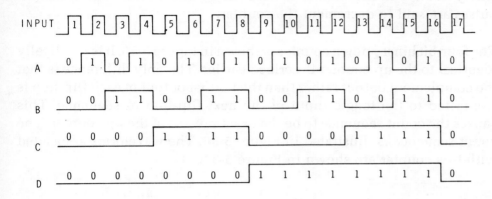

Figure 5-34
Input and output waveforms of a 4-bit binary counter.

In observing the waveforms in Figure 5-34, you should note several important things. First, all the flip-flops toggle (change state) on the trailing edge or the binary 1 to binary 0 transition of the previous flip-flop. With this in mind, you can readily trace the output of the first (A) flip-flop by simply observing when the trailing edges of the input occur.

The output of the B flip-flop is a function of its input, which is the output of the A flip-flop. Note that its state change occurs on the trailing edge of the A output. The same is true of the C and D flip-flops. The binary code after each input pulse is indicated on the waveforms. Of course, this corresponds to the binary count sequence in Figure 5-33.

Another important fact that is clear from the waveforms in Figure 5-34 is that the binary counter is also a frequency divider. The output of each flip-flop is one half the frequency of its input. If the input is a 100 kHz square wave, the outputs of the flip-flops are:

$$A — 50 \text{ kHz}$$
$$B — 25 \text{ kHz}$$
$$C — 12.5 \text{ kHz}$$
$$D — 6.25 \text{ kHz}$$

The output of a pure binary counter is always some sub-multiple of two. The four-bit counter divides the input by 16, (100 kHz ÷ 16 = 6.25 kHz).

DOWN COUNTER

The binary counter just described is referred to as an up-counter. Each time that an input pulse occurs, the binary number in the counter is increased by one. We say that the input pulses **increment** the counter. It is also possible to produce a down counter where the input pulses cause the binary number in the counter to decrease by one. In this case, the input pulses are said to **decrement** the counter.

The four-bit binary down counter is shown in Figure 5-35. It is practically identical to the up counter described earlier. The only difference is that the complement output rather than the normal output of each flip-flop is connected to the toggle input of the next flip-flop in sequence. This causes the count sequence to be the exact reverse of the up counter. The count sequence is illustrated in Figure 5-36. The waveforms associated with this counter are shown in Figure 5-37.

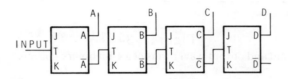

Figure 5-35
A 4-bit binary down counter.

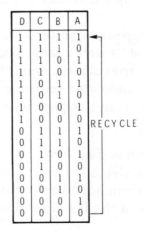

D	C	B	A
1	1	1	1
1	1	1	0
1	1	0	1
1	1	0	0
1	0	1	1
1	0	1	0
1	0	0	1
1	0	0	0
0	1	1	1
0	1	1	0
0	1	0	1
0	1	0	0
0	0	1	1
0	0	1	0
0	0	0	1
0	0	0	0

RECYCLE

Figure 5-36
Count sequence for 4-bit down counter.

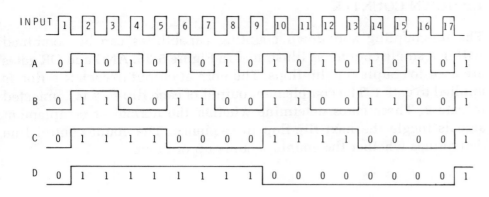

Figure 5-37
Input and output waveforms of a 4-bit binary down
counter.

In analyzing the operation of this counter, keep in mind that we still determine the contents of the counter by observing the normal outputs of the flip-flops as we did with the up-counter. Assuming the counter is initially reset and its contents are 0000, the application of an input pulse will cause all flip-flops to become set. With the A flip-flop reset, its complement output is high. When the first input pulse is applied, the A flip-flop will set. As it does, its complement output will switch from binary 1 to binary 0, thereby toggling the B flip-flop. The B flip-flop becomes set and its complement output also switches from binary 1 to binary 0. This causes the C flip-flop to set. In the same way, the complement output of the C flip-flop switches from high to low, thereby setting the D flip-flop. The counter recycles from 0000 to 1111.

When the next input pulse arrives, the A flip-flop will again be complemented. It will reset. As it resets, the complement output will switch from binary 0 to binary 1. The B flip-flop ignores this transition. No further state changes take place. The content of the counter then is 1110. As you can see, this input pulse causes the counter to be decremented from 15 to 14.

Applying another input pulse again complements the A flip-flop. It now sets. As it sets, its complement output switches from binary 1 to binary 0. This causes the B flip-flop to reset. As it resets, the complement output switches from binary 0 to binary 1. The C flip-flop ignores this transition. The new counter contents then is 1101, or 13. The input pulse again caused the counter to be decremented by one. By using the table in Figure 5-36 and the waveforms in Figure 5-37, you can trace the complete operation of the 4-bit binary down counter.

UP-DOWN COUNTER

The up-counting and down counting capabilities can be combined within a single counter as illustrated in Figure 5-38. AND and OR gates are used to couple the flip-flops. The normal output of each flip-flop is applied to gate 1. The complement output of each flip-flop is connected to gate 2. These gates determine whether the normal or complement signals toggle the next flip-flop in sequence. The count control line determines whether the counter counts up or down.

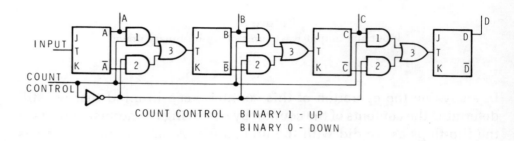

Figure 5-38
A binary up/down counter.

If the count control input is binary 1, all gate 1's are enabled. The normal output of each flip-flop then is coupled through gates 1 and 3 to the T input of the next flip-flop. The counter therefore counts up. During this time, all gate 2's are inhibited.

By making the count control line binary 0, all gate 2's are enabled. The complement output of each flip-flop is coupled through gates 2 and 3 to the next flip-flop in sequence. With this arrangement, the counter counts down.

BCD Counters

A BCD counter is a sequential circuit that counts by tens. It has ten discrete states, which represent the decimal numbers 0 through 9. Because of its ten-state nature, a BCD counter is also sometimes referred to as a decade counter.

The most commonly used BCD counter counts in the standard 8421 binary code. The table in Figure 5-39 shows the count sequence. Note that a four-bit number is required to represent the ten states 0 through 9. These ten four-bit codes are the first ten of the standard pure binary code. As count pulses are applied to the binary counter, the counter will be incremented as indicated in the table. Upon the application of a tenth input pulse, the counter will recycle from the 1001 (9) state to the 0000 state.

An 8421 BCD counter constructed with JK flip-flops is shown in Figure 5-40. This counter will generate the BCD code given in the Table of Figure 5-39. Note that the counter consists of four flip-flops like the four-bit pure binary counter discussed earlier. The output of one flip-flop drives the T input to the next in sequence. Unlike the binary counter discussed earlier, however, this circuit has several modifications which permit it to count in the standard 8421 BCD sequence. The differences consist of a feedback path from the complement output of the "D" flip-flop back to the J input of the "B" flip-flop. Also, a two input AND gate monitors the output states of flip-flops B and C and generates a control signal that is used to operate the J input to the D flip-flop. These circuit modifications, in effect, trick the standard four-bit counter and cause it to recycle every ten input pulses.

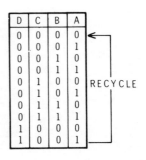

D	C	B	A
0	0	0	0
0	0	0	1
0	0	1	0
0	0	1	1
0	1	0	0
0	1	0	1
0	1	1	0
0	1	1	1
1	0	0	0
1	0	0	1

RECYCLE

Figure 5-39
Count sequence of
8421 BCD counter.

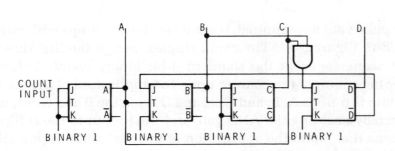

Figure 5-40
An 8421 BCD counter.

The waveforms shown in Figure 5-41 illustrate the operation of the 8421 BCD counter. The count sequence is identical to that of the standard four-bit pure binary counter discussed earlier for the first eight input pulses. The operations that occur during the 9th and 10th pulses are unique to the BCD counter.

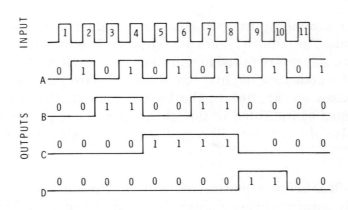

Figure 5-41
Waveforms of the 8421 BCD counter.

Assume that the counter in Figure 5-40 is initially reset. The outputs of flip-flops B and C will be binary 0 at this time. This makes the output of the AND gate low and causes the J input of the D flip-flop to be held low. The D flip-flop, cannot be set by the toggle input from the A flip-flop until the J input goes high. Note also that the complement output of the D flip-flop, which is binary 1 during the reset state, is applied to the J input of the B flip-flop. This enables the B flip-flop, permitting it to toggle when the A flip-flop changes state.

If count pulses are now applied, the states of the flip-flops will change as indicated in Figure 5-41. The count ripples though the first three flip-flops in sequence as in the standard 4-bit binary counter. However, consider the action of the counter upon the application of the 8th input pulse. With flip-flops A, B, and C set and D reset, the B and C outputs are high, thereby enabling the AND gate and the J input to the D flip-flop. This means that, upon the application of the next count input, all flip-flops will change state. The A, B, and C flip-flops will be reset while the D flip-flop is set. The counter state changes from 0111 to 1000 when the trailing edge of the 8th input pulse occurs.

In this new state, the B and C outputs are low, therefore causing the J input to the D flip-flop to again be binary 0. With the J input 0 and the K input binary 1 and the D flip-flop set, the conditions are right for this flip-flop to be reset when the T input switches from binary 1 to binary 0. In additon, the complement output of the D flip-flop is low at this time, thereby keeping the J input to the B flip-flop low. The B flip-flop is reset at this time and, therefore, the occurance of a clock pulse at the T input will not affect the B flip-flop.

When the 9th input pulse occurs, the A flip-flop sets. No other state changes occur at this time. The binary number in the counter is now 1001. The transition of the A flip-flop switching from binary 0 to binary 1 is ignored by the T input of the D flip-flop.

When the 10th input pulse occurs, the A flip-flop will toggle and reset. The B flip-flop will not be affected at this time since its J input is low. No state change occurs in the C flip-flop since the B flip-flop remains reset. The changing of the state of the A flip-flop, however, does cause the D flip-fop to reset. With its J input binary 0 and K input binary 1, this flip-flop will reset when the A flip-flop changes state. This 10th input pulse therefore causes all flip-flops to become reset. As you can see by the waveforms in Figure 5-41, the counter recycles from the 1001 (9) state to the 0000 state on the 10th input pulse.

Like any counter, the BCD counter can also be used as a frequency divider. Since the BCD counter has ten discrete states, it will divide the input frequency by ten. The output of the most significant bit flip-flop in the BCD counter will be one tenth of the input frequency. From Figure 5-40, you can see that only a single output pulse occurs at the D output for every ten input pulses. While the D output does not have a 50 percent duty cycle, the frequency of the signal is nevertheless one tenth of the input frequency.

Shift Registers

The illustration in Figure 5-42 shows you how a shift register operates. Here, the shift register consists of four binary storage elements, such as flip-flops. The binary number 1011 is currently stored in the shift register. Another binary word, 0110, is generated externally. As shift pulses are applied, the number stored in the register will be shifted out and lost while the external number would be shifted into the register and retained.

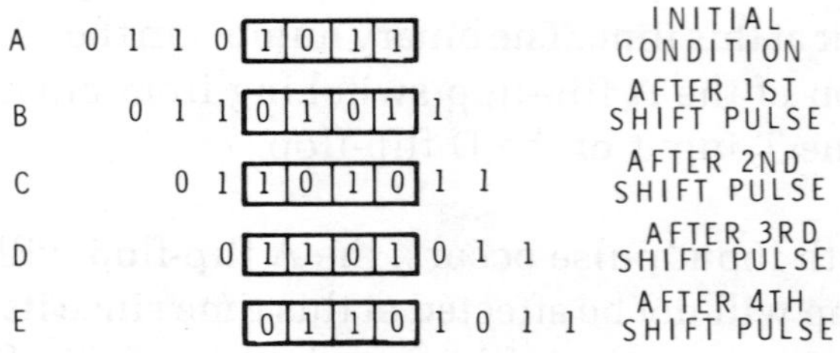

```
A    0 1 1 0 [1 0 1 1]              INITIAL
                                   CONDITION
B      0 1 1 [0 1 0 1] 1          AFTER 1ST
                                  SHIFT PULSE
C        0 1 [1 0 1 0] 1 1        AFTER 2ND
                                  SHIFT PULSE
D          0 [1 1 0 1] 0 1 1      AFTER 3RD
                                  SHIFT PULSE
E            [0 1 1 0] 1 0 1 1    AFTER 4TH
                                  SHIFT PULSE
```

Figure 5-42
Operation of a shift register.

The initial conditions for this shift register are illustrated in Figure 5-42A. After one clock pulse, the number stored in the register initially is shifted one bit position to the right. The right-most bit is shifted out and lost. At the same time, the first bit of the externally generated number is shifted into the left-most position of the shift register, this is illustrated in Figure 5-42B. The remaining three illustations in C, D, and E show the results after the application of additional shift pulses. After four shift pulses have occurred, the number originally stored in the register has been completely shifted out and lost. The number appearing at the input on the left has been shifted into the register and now resides there.

An important point to note is that the data is shifted one bit position for each input clock, or shift, pulse. Clock pulses have full control over the shift register operation. In this shift register, the data was shifted to the right. However, in other shift registers, it is also possible to shift data to the left. The direction of the shift is determined by the application. Most shift registers are of the shift-right type.

The shift register is one of the most versatile of all sequential logic circuits. It is basically a storage element used for storing binary data. A single shift register made up of many storage elements can be used as a memory for storing many words of binary data.

Shift registers can also be used to perform arithmetic operations; shifting the data stored in a shift register to the right or to the left a number of bit positions is equivalent to multiplying or dividing that number by a specific factor. They can also be used to generate a sequence of control pulses for a logic circuit. And in some applications, shift registers can be used for counting and frequency dividing.

Shift registers are usually implemented with JK flip-flops. Type D flip-flops can also be used, but shift registers implemented with JK flip-flops are far more versatile. A typical shift register constructed with JK flip-flops is shown in Figure 5-43. The input data and its complement are applied to the JK inputs of the input (A) flip-flop. From there, the other flip-flops are cascaded with the outputs of one connected to the JK inputs of the next. Note that the clock (T) input lines to all flip-flops are connected together. The clock or shift pulses are applied to this line. Data applied to the input will be shifted to the right through the flip-flops. Each clock, or shift, pulse will cause the data at the input and that stored in the flip-flops to be shifted one bit position to the right.

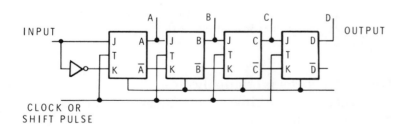

Figure 5-43
A 4-bit shift register made with JK flip-flops.

The waveforms in Figure 5-44 illustrate how a data word is loaded into the shift register of Figure 5-43. As the waveforms show, the binary number 0101 occurs in synchronization with the input clock, or shift, pulses. In observing the waveforms in Figure 5-44, keep in mind that time moves from left to right. This means that the clock pulses on the right occur after those on the left. In the same way, the state of the input shown on the left occurs prior to the states to the right. With this in mind, let's see how the circuit operates.

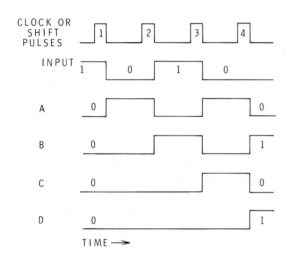

Figure 5-44
Waveforms illustrating how the serial binary number
0101 is loaded into a shift register.

Note that the shift register is originally reset. The A, B, C, and D outputs of the flip-flops, therefore, are binary 0 as indicated in the waveforms. Prior to the application of the number 1 shift pulse, the input state is binary 1. This represents the first bit of the binary word to be entered. On the trailing edge of the first clock pulse, the binary one will be loaded into the A flip-flop. The JK inputs of the A flip-flop are such that when the clock pulse occurs the flip-flop will become set. This first shift pulse is also applied to all other flip-flops. The state stored in the A flip-flop will be transferred to the B flip-flop. The states stored in the B and C flip-flops will be transferred to the C and D flip-flops respectively. Since all flip-flop states are initially zero, no state changes in the B, C, or D flip-flops will take place when the first clock pulse occurs.

After the first clock pulse, the A flip-flop is set while the B, C, and D flip-flops are still reset. The first clock pulse also causes the input word to change. The clock, or shift, pulses are generally common to all other circuits in the system; therefore, any data available in the system will generally be synchronized to the clock.

The input to the A flip-flop is now binary 0. When the trailing edge of the second clock pulse occurs, this binary 0 will be written into the A flip-flop. The A flip-flop, which was set by the first clock pulse, causes the JK inputs to the B flip-flop to be such that it will become set when the second clock pulse occurs. As you can see by the waveforms, when the second clock pulse occurs, the A flip-flop will reset while the B flip-flop will set. The 0 state previously stored in the B flip-flop will be transferred to the C flip-flop, and the C flip-flop state will be shifted to the D flip-flop. At this point the first two bits of the data word have been loaded into the shift register.

The input is now binary 1, representing the third bit of the input word. When the third clock pulse occurs, the A flip-flop will set. The zero previously stored in the A flip-flop will be transferred to the B flip-flop. The binary 1 stored in the B flip-flop will now be shifted into the C flip-flop. The D flip-flop remains reset.

The input to the A flip-flop is now binary 0. When the trailing edge of the fourth clock pulse occurs, the A flip-flop will reset. The binary 1 stored there previously will be transferred to the B flip-flop. The 0 stored in the B flip-flop will be shifted into the C flip-flop. The binary 1 in the C flip-flop now moves to the D flip-flop. As you can see, after four clock pulses have occurred, the complete four-bit binary word 0101 is now shifted into the register, as indicated by the states shown in the waveforms. A glance at the flip-flop output waveforms will show the initial binary 1 bit moving to the right with the occurrrence of each shift pulse.

While we have illustrated the operation of the shift registers with only four bits. Naturally, as many flip-flops as needed can be cascaded to form longer shift registers. Most shift registers are made up to store a single binary word. In most modern digital systems, shift registers have a number of bits that is some multiple of four.

Self Review Questions

30. Name the two most widely used types of sequential logic circuits.
    ⎯⎯⎯⎯⎯⎯⎯ ⎯⎯⎯⎯⎯⎯⎯

31. The circuit of Figure 5-45 is a ⎯⎯⎯⎯-bit ⎯⎯⎯⎯⎯⎯⎯⎯⎯
    ⎯⎯⎯⎯⎯⎯⎯⎯⎯ counter.
    up/down

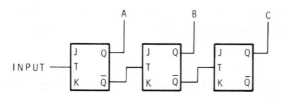

Figure 5-45
Identify this circuit.

32. The counter of Figure 5-45 contains the number 010. Six input pulses occur. The new counter state is ⎯⎯⎯⎯⎯⎯⎯⎯.

33. A binary counter made up of five JK flip-flops will divide an input frequency by ⎯⎯⎯⎯⎯⎯⎯⎯.

34. A BCD counter contains the number 1000. Six input pulses occur. The new counter state is _____.

35. A BCD counter divides its input signal frequency by _____.

36. A BCD counter is cascaded with a three-flip-flop binary counter. The overall frequency division ratio is _____.

37. What is a shift register? _____
   _____
   _____

38. An 8-bit shift register contains the number 10000110. The number 11011011 is applied to the input. After five shift pulses, what is the number in the shift register? (Assume shift right operation.)
   _____

39. How many shift pulses are required to load a 16-bit word into a 16-flip-flop shift register? _____

Self Review Answers

30. The two most widely used types of sequential logic circuits are the **counter** and the **shift register**.

31. The circuit of Figure 5-45 is a **3-bit binary down counter**.

32. The new counter state is 000. The counter starts at 010 or decimal 2. After five input pulses, the counter will be at 111, or decimal 7. Therefore, the sixth input pulse will cause the counter to reset to 000.

33. A binary counter made up of five JK flip-flops will divide an input frequency by 32.

34. The new counter state is 0100. The counter starts at 1000, or decimal 8. After one input pulse, the counter will be at 1001, the maximum count of a BCD counter. After the second input pulse, the counter will reset to 0000. Now the counter will continue its count for the four remaining pulses to 0100, or decimal 4.

35. A BCD counter divides its input signal frequency by **10**.

36. The overall frequency division ratio is 80. The BCD counter divides by 10. The three flip-flop binary counter divides by 8. The total division is the product of the two, or $10 \times 8 = 80$.

37. A shift register is basically a storage element used for storing binary data.

38. The number in the shift register after five shift pulses is 11011100.

```
        INPUT        ORIGINAL REGISTER CONTENT
    11011011 ──▶  │ 1  0  0  0  0  1  1  0 │

                         AFTER FIVE SHIFT PULSES
        110 ──▶  │ 1  1  0  1  1  0  0 │ 0  0  1  1  0
```

39. Sixteen shift pulses are required to load a 16-bit word into a 16-flip-flop shift register.

CLOCKS AND ONE SHOTS

Most sequential logic circuits are driven by a clock, a periodic signal that causes logic circuits to be stepped from one state to the next through their normal operating states. The clock signal is generated by a clock oscillator circuit, which generates rectangular output pulses with a specific frequency, duty cycle, and amplitude.

Practically all digital clock oscillator circuits use some form of **astable** circuit to generate these periodic pulse waveforms. Such a circuit switches repeatedly between its two **un**stable states, hence the name astable.

Another circuit that is widely used, in addition to the clock, to implement sequential logic operations is known as the "one shot multivibrator," or "one shot." This circuit produces a fixed duration output pulse each time it receives an input trigger pulse. Its normal state is "off", until it receives a trigger pulse. Therefore, it is called a **monostable** circuit, since it has one stable state. The duration of its output pulse is usually controlled by external components.

This section discusses the very widely used 555 timer integrated circuit. It can be used either as a clock oscillator or as a one shot.

The 555 Timer

The 555 timer is a low cost linear IC which has dozens of different functions. It can act as a monostable circuit or as an astable circuit, and it has many other applications as well. However, this discussion will be limited to its astable and monostable modes of operation. For a more detailed discussion of the 555 and other IC timers, see the Heath "Electronics Technology Series" course number EE-103, "IC Timers."

Many different manufacturers produce the 555 Timer. Different versions have numbers like SE555, CA555, SN72555, and MC14555. However, you will notice that all contain the basic 555 number. Dual 555 timers are also available on a single chip and they carry "556" numbers.

A simplified diagram of the 555 circuit is shown in Figure 5-46. Notice that it contains two comparators, a flip-flop, an output stage, and a discharge transistor (Q_1). With the proper external components, several different functions can be implemented.

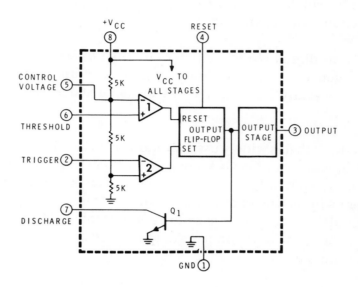

Figure 5-46
The 555 timer circuit.

If you examine the comparators in more detail, you will see that they are actually op amps with no feedback path. Therefore, they have an extremely high voltage gain. In fact, a few microvolts input will cause the output to drive to the full supply voltage or to zero volts, depending upon the input polarity.

In this circuit, a voltage divider consisting of three 5Ω resistors develops a reference voltage at one input of each comparator. The reference voltage at the — input of comparator 1 is 2/3 of V_{cc}. The other input to the comparator comes from an external circuit via pin 6. When the voltage at pin 6 rises above the reference voltage, the output of comparator 1 swings positive. This, in turn, resets the flip-flop.

The reference voltage at the + input of comparator 2 is set by the voltage divider at 1/3 of V_{cc}. The other input to comparator 2 is the trigger input. When the trigger input falls below the reference voltage, the output of the comparator swings positive. This sets the flip-flop.

The output of the flip-flop will always be at one of two levels. When the flip-flop is reset, its output goes to a positive voltage which we will call +V. When set, its output falls to a very low voltage which we will call 0 volts.

The output of the flip-flop is amplified and inverted by the output stage. A load can be connected between the output terminal (pin 3) and either $+V_{cc}$ or ground. When the load is connected to $+V_{cc}$, a heavy current flows through the load when the output terminal is at 0 volts. Little current flows when the output is at +V. However, if the load is connected to ground, maximum current flows when the output is at +V and little current flows when the output is at 0 volts.

Notice that the output of the flip-flop is also applied to the base of Q_1. When the flip-flop is reset, this voltage is positive and Q_1 acts as a very low impedance between pin 7 and ground. On the other hand, when the flip-flop is set, the base of Q_1 is held at 0 volts. Thus, Q_1 acts as a high impedance between pin 7 and ground.

The 555 One Shot

Figure 5-47A shows the 555 timer being used as a one shot or monostable. This circuit produces one positive pulse output for each negative pulse at the trigger input. The duration of the output pulse can be precisely controlled by the value of external components C_1 and R_A.

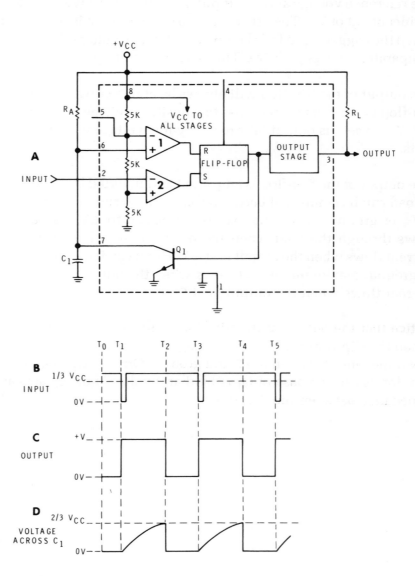

Figure 5-47
The 555 timer as a monostable circuit.

Figure 5-47B shows the input pulses. Between pulses, the input voltage is held above the trigger voltage of comparator 2. The flip-flop is reset and its output is at +V. This output is inverted by the output stage so pin 3 is at 0 volts. Thus, a heavy current flows through R_L. The output of the flip-flop (+V) is also applied to the base of Q_1, causing the transistor to conduct. Q_1 acts as a short across C_1. These conditions are shown at time T_0. Notice that the output voltage (Figure 5-47C) and the capacitor voltage (Figure 5-47D) are at 0 volts at this time.

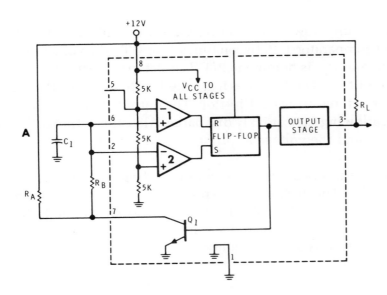

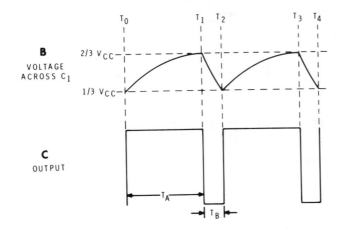

Figure 5-48
The 555 timer as an astable circuit.

At time T_1, a negative input pulse occurs. This forces the voltage at the − input of comparator 2 below the 1/3 V_{cc} reference. The comparator switches states, setting the flip-flop. The output of the flip-flop falls to 0 volts. This voltage is applied to Q_1, cutting the transistor off. This removes the short from around C_1 and the capacitor begins to charge through R_4 toward $+V_{cc}$.

The output of the flip-flop is also applied to the output stage where it is inverted. Thus, the output at pin 3 swings to $+V$. The output will remain in this state until the flip-flop is reset.

In the circuit shown, you can only reset the flip-flop by switching the state of comparator 1. Between times T_1 and T_2, the voltage at the + input of comparator 1 is below the 2/3 V_{cc} reference. Capacitor C_1 is charging toward this level and reaches it at time T_2. This switches the output of the comparator, resetting the flip-flop. The output of the flip-flop turns on Q_1 again, allowing C_1 to quickly discharge. Also, the output of the flip-flop is inverted and the voltage at pin 3 falls back to 0 volts.

The output is a positive pulse whose leading edge is determined by the input pulse. The duration of the pulse is determined by the time required for C_1 to charge to 2/3 of V_{cc}. This, in turn, is determined by the $R_4 C_1$ time constant. C_1 can charge to 2/3 V_{cc} in just over one time constant. Thus the pulse duration is approximately:

$$PD = 1.1\ R_4 C_1$$

The charge rate of C_1 and the threshold voltage of comparator 2 are both directly proportional to $+V_{cc}$. Thus, the pulse duration remains virtually constant regardless of the value of $+V_{cc}$.

If C_1 is a 0.01 μF capacitor and R_1 is a 1 megohm resistor, the pulse duration is

$$PD = 1.1\ (1 \times 10^6\ \Omega)\ (1 \times 10^{-8}\text{F})$$
$$PD = 1.1\ (1 \times 10^{-2})$$
$$PD = 1.1 \times 10^{-2} \text{ seconds or 11 milliseconds.}$$

The 555 Clock

Another application of the 555 timer is shown in Figure 5-48A. This is an astable circuit. It free runs at a frequency determined by C_1, R_A, and R_B. Figure 5-48B and C show the voltage across C_1 and the output voltage. Between times T_0 and T_1, the flip-flop is set and its output is 0 volts. This holds Q_1 cut off and holds the output (pin 3) at $+V$.

With Q_1 cut off, C_1 begins to charge toward $+V_{cc}$ through R_B and R_A. At time T_1, the voltage across the capacitor reaches 2/3 of $+V_{cc}$. This causes the flip-flop to reset. The output of the flip-flop goes to $+V$ and the output at pin 3 drops to 0 volts. Q_1 conducts allowing C_1 to discharge through R_B. As C_1 discharges, the voltage across C_1 decreases. At time T_2, the voltage has decreased to the trigger level of comparator 2. This sets the flip-flop again, cutting off Q_1. The capacitor begins to charge once more and the entire cycle is repeated.

As you can see, the capacitor charges and discharges between 2/3 of $+V_{cc}$ and 1/3 of $+V_{cc}$. C_1 charges through both R_A and R_B. Approximately 0.7 time constants are required for C_1 to charge. Thus, the duration of the positive output pulse (T_A) is approximately

$$T_A = 0.7\, C_1\, (R_A + R_B)$$

Also, the duration of the negative going pulse is determined by the $C_1 R_B$ time constant. Consequently,

$$T_B = 0.7\, C_1 R_B$$

The total period of one cycle is

$$T = T_A + T_B$$

And, since frequency is the reciprocal of time,

$$f = \frac{1}{T} = \frac{1}{T_A + T_B} = \frac{1.43}{C_1\, (R_A + 2\, R_B)}$$

For example, if C_1 is a 0.01 μF capacitor, R_A is a 1 MΩ resistor, and R_B is a 100 kΩ resistor, the positive output pulse duration is:

$$T_A = 0.7\ C_1\ (R_A + R_B)$$

$$= 0.7\ (0.01\ \mu F)\ (1\ M\Omega + 100\ k\Omega)$$

$$= 0.7\ (1 \times 10^{-8}\ F)\ (1.1 \times 10^6\ \Omega)$$

$$= 7.7 \times 10^{-3}\ \text{seconds or 7.7 milliseconds}$$

The duration of the negative going output pulse is:

$$T_B = 0.7\ C_1\ R_B$$

$$= 0.7\ (0.01\ \mu F)\ (100\ k\Omega)$$

$$= 0.7\ (1 \times 10^{-8}\ F)\ (1 \times 10^5\ \Omega)$$

$$= 7 \times 10^{-4}\ \text{seconds or 0.7 milliseconds}$$

The total period of one cycle is then:

$$T = T_A + T_B$$

$$T = 7.7\ \text{milliseconds} + 0.7\ \text{milliseconds}$$

$$T = 8.4\ \text{milliseconds}$$

And the frequency is:

$$f = \frac{1}{T}$$

$$f = \frac{1}{8.4\ \text{milliseconds}}$$

$$f = 119\ \text{Hz}$$

Self-Review Questions

40. What is an astable circuit? _____
_____
_____

41. What is a monostable circuit? _____
_____
_____

42. In a 555 monostable circuit, if R_A is 2 kΩ and C_1 is 1 μF, what is the output pulse duration?

43. In a 555 astable circuit, if R_A is 100 kΩ, R_B is 500 kΩ and C_1 is 0.1 μF, what is the output frequency?

Self-Review Answers

40. An astable is a circuit that has two unstable states. It switches between these two states continuously, thus generating a periodic pulse waveform.

41. A monostable circuit has one stable state. When it is triggered, it produces a fixed duration output pulse and then returns to its stable state.

42. $PD = 1.1\ R_A C_1$

    $PD = 1.1\ (2\ k\Omega)\ (1\ \mu F)$

    $PD = 1.1\ (2 \times 10^3\ \Omega)\ (1 \times 10^{-6}\ F)$

    $PD = 2.2 \times 10^{-3}$ seconds

    $PD = 2.2$ milliseconds

43.

$$f = \frac{1.43}{C_1\ (R_A + 2\ R_B)}$$

$$f = \frac{1.43}{0.1\ \mu F\ [\ 100\ k\Omega + 2\ (500\ k\Omega)\ ]}$$

$$f = \frac{1.43}{(0.1 + 10^{-6}\ F)\ (1.1 \times 10_6\ \Omega)}$$

$$f = \frac{1.43}{0.11}$$

$$f = 13\ Hz$$

Unit 6

DIGITAL COMPUTERS

CONTENTS

INTRODUCTION

Digital circuits were originally developed to provide a means of implementing digital computers. As new circuits and techniques were developed, computer performance was improved. But the greatest impact on digital computers has been the development of integrated circuits, or "IC's," which made them more powerful and greatly reduced their size and cost. Over the years, digital computers have continued to decrease in price. Their size and power consumption have also decreased significantly. At the same time, their performance and sophistication have increased, making them practical for a wider range of applications.

Recent technological advances in semiconductor techniques have created a unique digital product. Large scale integration of digital circuits have permitted the semiconductor manufacturers to put an entire digital computer on a single chip of silicon. These computers are known as microprocessors. We normally think of digital integrated circuits as being the gates and flip-flops used to implement a computer. Now, the computer itself is a single, low-cost integrated circuit. But the power of this device is significant, and for many applications it can replace hundreds of small scale and medium scale integrated circuits. This significant development will further broaden the applications for digital computers. Best of all, it will increase the sophistication and capabilities of the electronic equipment that uses them.

While it is impossible to cover all aspects of this exciting field in this unit, it will introduce you to the digital computer and its related techniques. The primary emphasis will be on the microprocessor and its ability to replace standard hard-wired digital logic systems.

The "Unit Objectives" are listed next. They state exactly what you are expected to learn in this unit. Review this list now and refer to it as you are completing this unit to be sure you meet each objective.

UNIT OBJECTIVES

When you have completed this unit you will be able to:

1. Define digital computer, data, instructions, program, software, and peripheral unit.

2. Name the four types of small computers.

3. List the four major sections of a digital computer and define each.

4. Define computer instruction, address, central processing unit, accumulator, and interrupt.

5. List the seven steps of computer programming.

6. Define machine language programming, algorithm, flow chart, coding, and loop.

7. Analyze a simple program, when given the computer instruction set.

8. Define subroutine, assembler, compiler, cross-assembler, cross-compiler, and utility program.

9. State the primary use and applications of the microprocessor.

10. List the benefits of using a microprocessor over a hard-wired logic system.

11. Write a simple microprocessor program when given the problem and computer instruction set.

WHAT IS A DIGITAL COMPUTER?

A **digital computer** is an electronic machine that automatically processes data by the use of digital techniques. **Data** refers to any information such as numbers, letters, words, or even complete sentences and paragraphs. Processing is a general term referring to a variety of ways in which the data can be manipulated. The computer processes the data by performing arithmetic operations on it, editing and sorting it, or evaluating its characteristics and making decisions based upon it. In addition to being able to manipulate data in a variety of ways, the computer contains an extensive memory where data is stored. The key characteristic of a digital computer is its ability to process data automatically without operator intervention.

The manner in which the data is manipulated is determined by a set of instructions, or "instruction set," contained within the machine. These instructions form a **program** that tells the computer exactly how to handle the data. The instructions are executed sequentially to carry out the desired manipulations. Most computers are general purpose, in that the instructions can be assembled into an almost infinite variety of application programs.

Each computer has a specific instruction set. These instructions are put into the proper sequence so they will perform the required calculation or operation. The process of writing the desired sequence of instructions is called programming.

How Computers are Classified

There are many different types of digital computers and a variety of ways in which they can be classified. One method of classifying computers is by size and computing power. At one end of the spectrum are large-scale computers with extensive memory and high-speed calculating capabilities. These machines can process huge volumes of data in a short period of time and in any desired manner. At the other end of the spectrum are the small-scale, low-cost digital computers, such as the microprocessor — whose application and computing power is more limited.

Computers are also classified by function or application. The most commonly known digital computer is the electronic data processor that is used by most business industry, and government organizations to maintain records, perform accounting functions, maintain an inventory, and provide a wide variety of other data processing functions. Then there are the scientific and engineering computers that are used primarily as mathematical problem solvers. They greatly speed up and simplify the calculations of complex and difficult scientific and engineering problems.

Another way to classify digital computers is general purpose or special purpose. General purpose machines are designed to be as flexible as possible. This means that they can be programmed for virtually any application. Special purpose computers, on the other hand, are generally dedicated to a specific application. They are designed to carry out only a single function. General purpose computers with a fixed program become special purpose computers.

Most digital computers are of the general purpose type, and most have versatile instruction sets so they can be programmed to perform almost any operation. With the proper program, a general purpose computer can perform business data processing functions, scientific and mathematical calculations, or industrial control functions.

The most widely used computers are the small-scale machines. These include the minicomputer, the microcomputer, the programmable calculator, and the microprocessor. While all of these small-scale machines together account for less than 10% of the total computer dollar investment, they represent more than 95% of the unit volume of computers. Small-scale computer systems are very low priced. Today, you can purchase a complete computer system for less than the price of a new automobile. Microprocessors and programmable calculators are even less expensive. There are many thousands of small computers in use today. Your own personal contact with a digital computer will no doubt be through some type of small-scale computer.

MINICOMPUTERS

The minicomputer is the largest of the four types of small computers. This is a general purpose digital computer, usually constructed of bipolar logic circuits, and supported with software and peripheral units. **Software** refers to the programs supplied with the computer that make it easy to use. **Peripheral units** are the input-output devices that allow an operator to communicate with the computer. Typical peripheral units are typewriters, card readers and printers. Minicomputers are similar to the larger digital computers, but their memory capacity, speed, and applications are more limited.

You can purchase a complete but minimum minicomputer for less than $1,000. This does not include peripheral equipment. However, such machines are often purchased to be built into a larger piece of equipment or a system for use as a controller. The users of such computers are referred to as original equipment manufacturers (OEM). A complete stand-alone minicomputer with sufficient memory, peripheral devices, and software to be used for general purpose computing may cost less than $5,000.

MICROCOMPUTERS

A microcomputer is similar in many respects to a minicomputer in that it is a general purpose machine that can be programmed to perform a wide variety of functions. However, the microcomputer is normally smaller and more restricted in its application. Its speed and memory capacity is less than a minicomputer. As a result, microcomputers are substantially less expensive than minicomputers. Microcomputers are more often used in dedicated, single function applications. Software and peripheral support is minimum. Most microcomputers are implemented with MOS LSI circuitry.

PROGRAMMABLE CALCULATOR

A programmable calculator can be classified as a special purpose microcomputer. These machines are similar in many respects to hand-held and desk-top electronic calculators. The programmable calculator has an input keyboard for entering data and a decimal display for reading out the results of calculations.

In a standard calculator, an operator enters the numbers to be manipulated and the functions to be performed by depressing keys on the keyboard in the proper sequence. The solution to the problems then appear on the display. A programmable calculator can also be used in this way, but it contains a memory and control unit that is used to automate the problem solving process. The data to be operated upon and the functions to be performed are entered via the keyboard and stored in the memory in the proper sequence. When it is enabled, the programmable calculator will then automatically solve the problem stored in its memory without operator control.

Programmable calculators offer the advantage of improved speed and convenience over standard calculators when the same problem must be computed several times with different data. Long problems requiring complex data and many mathematical operations are also best solved by a programmable calculator, as they relieve the operator from the tedious work and greatly minimize errors. Another advantage of the programmable calculator over other types of digital computers is its ability to communicate directly with the operator through the keyboard and decimal readout display.

MICROPROCESSORS

A microprocessor is the smallest and least expensive type of digital computer that still retains all of the basic features and characteristics of a computer. It can be implemented with standard digital integrated circuits or it is available as a single large scale integrated (LSI) circuit. While the capabilities of a microprocessor are limited when compared with a microcomputer or minicomputer, this device is still a very powerful unit. It extends the applications of computer techniques to many areas where minicomputers and microcomputers are not economically feasible.

Microprocessors are generally designed to perform a dedicated function. These devices are built into electronic equipment that will be used for some specific application. Some typical dedicated applications include traffic light controllers, electronic scales and cash registers, and electronic games. In addition, engineers are finding that low cost microprocessors can be used to replace standard hard-wired digital logic. Design time and cost can be significantly reduced in the design of a digital system when microprocessors are used.

A microprocessor can be used economically if the design is equivalent to thirty or more standard integrated circuit packages. Such hard wired logic designs are replaced by a microprocessor with a stored program. The program stored in a read only memory permits the microprocessor to carry out the same functions as a hard-wired logic controller. Microprocessors can also be used as the main component of a minicomputer or microcomputer.

Self Review Questions

1. What is a digital computer? _____
   _____
   _____

2. What is data? _____
   _____
   _____

3. The way the computer manipulates data is determined by a set of
   _____.

4. A list of computer instructions for solving a particular problem is
   called a _____.

5. What is software? _____
   _____
   _____

6. List several types of computer peripheral units. _____
   _____
   _____
   _____

7. Name the four types of small computers. _____
   _____
   _____

Self-Review Answers

1. A digital computer is an electronic machine that uses digital techniques to automatically process data.

2. Data is any information such as numbers, letters, words, or even complete sentences and paragraphs.

3. The way the computer manipulates data is determined by a set of **instructions.**

4. A list of computer instructions for solving a particular problem is called a **program**.

5. Software refers to the programs supplied with a computer.

6. Peripheral units are typewriters, card readers, printers, and other input-output devices.

7. The four types of small computers are:

> minicomputers
> microcomputers
> programmable calculators
> microprocessors

DIGITAL COMPUTER ORGANIZATION AND OPERATION

All digital computers are made up of four basic units: the memory, the control unit, the arithmetic logic unit (ALU), and the input-output (I/O) unit. These major sections and their relationship to one another are illustrated in Figure 6-1. An understanding of digital computer operation starts with a knowledge of how these sections operate and how they affect one another.

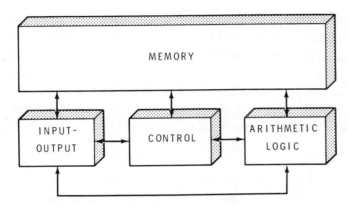

Figure 6-1
General block diagram of a digital
computer.

Memory

The heart of any digital computer is its memory, where the program and data are stored. As indicated earlier, the program is a series of instructions that are stored and executed in sequence to carry out some specific function. The instructions cause the computer to manipulate the data in some way.

Computer memories are organized as a large group of storage locations for fixed length binary words. A computer instruction is nothing more than a binary word whose bit pattern defines a specific function to be performed. The data to be processed by the computer is also a binary word. A computer memory is an accumulation of storage registers for these instruction and data words. Most computers have memories capable of storing many thousands of words.

Digital computers typically have a fixed word size. A 32-bit word is common for many large computers. Minicomputers usually have a 16-bit word. Microprocessors widely use an 8-bit word. Memory sizes range from approximately several hundred words to several hundred thousand words of storage. A typical minicomputer may provide 4096 16-bit words. A microprocessor may use 1024 words of 8-bit memory. The number of words in memory is generally some power of two.

Each memory location appears to be like a storage register. Data can be loaded into the register and retained. The word can also be read out of memory for use in performing some operation.

Each memory word is given a numbered location called an **address**. The address is a binary word used to locate a particular word in memory. The normal procedure is to store the instruction words in sequential memory locations. The instruction word generally contains an address which refers to the location of some data word to be used in carrying out the operation specified. The instructions stored in the sequential memory locations are executed one at a time until the desired function is performed.

Most modern digital computers use semiconductor memory, MOS LSI circuits where data is stored in latch flip-flops or as the charge on a capacitor. Semiconductor memories are small, fast, and inexpensive. Many computers, however, still use magnetic core memories. In these memories, binary data is stored in tiny donut-shaped magnetic cores. By magnetizing the core in one direction a binary zero is stored. Magnetizing the core in the opposite direction causes it to store a binary one. Electronic circuitry associated with the cores is used to store data into the memory and read it out.

The advantage of core memories over semiconductor memories is their non-volatility. When power is removed from a semiconductor memory, all of the data is lost. Removing the power from a magnetic core memory has no effect on the data contents. Because the cores are permanently magnetized in one direction or the other, all data is retained.

The typical organization of a computer memory is shown in Figure 6-2. It consists of the semiconductor or magnetic core elements that retain the binary data. The memory used in a digital computer is generally referred to as a random access read/write memory. Random access refers to the ability of the computer to directly seek out and access any specific word stored in the computer memory. Read/write refers to the ability of the memory to store data (write) or to retrieve data for use elsewhere (read).

As you can see from Figure 6-2, the access to a specific word in memory is achieved through the memory address register (MAR) and memory address decoder. The memory address register is a flip-flop register into which is placed a multi-bit binary word that designates the location of a desired word in memory. If the address 0001 0011 is stored in the MAR, the content of memory location 19 is referenced. The address word may refer to the location of an instruction or a data word. The size of the address word determines the maximum memory size. For example, if the memory address word is 12-bits in length, the maximum number of words that the computer memory can contain is $2^{12} = 4096$ words (called a 4K memory).

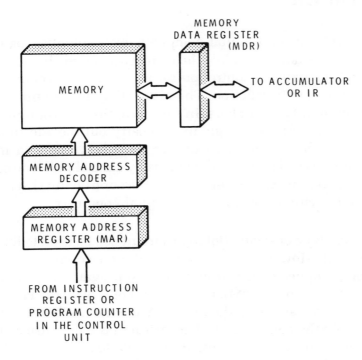

Figure 6-2
Typical computer memory organization.

The output of the memory address register drives the memory address decoder, which recognizes one unique memory address word at a time and enables the appropriate location. In semiconductor memories, the memory address decoder is generally a fixed part of the integrated circuit memory itself. When an address word is loaded into the MAR, the specific location in memory designated by that address is enabled. Data can then be written into or read out of that memory location.

The access to the addressed memory location is made through a memory data register (MDR) or memory buffer register (MBR). This is a flip-flop register into which the data or instruction word is stored on its way into or out of the memory. A word to be stored in memory is first loaded into the MDR and then stored in the addressed memory location. If a read operation is being carried out by the memory, the data stored in the addressed location is first loaded into the MDR. From there it is set to other portions of the computer as needed. Many computers do not use an MDR. Instead, the data or instruction goes to or comes from another register in the computer.

Control Unit

The control unit in a digital computer is a sequential logic circuit. Its purpose is to examine each of the instruction words in memory, one at a time, and generate the control pulses necessary to carry out the function specified by that instruction. The instruction, for example, may call for the addition of two numbers. In this case, the control unit would send pulses to the arithmetic logic unit to carry out the addition of the two numbers. If the instruction calls for the storage data in memory, the control unit would generate the necessary control pulses to carry out that storage operation. As you can see, it is the control unit that is responsible for the automatic operation of the digital computer.

Almost any type of sequential logic circuit can be used to implement the control unit. However, most modern digital computers incorporate a microprogrammed control unit using a preprogrammed circuit known as a read only memory (ROM). Here, special binary words known as micro-instructions are stored in the read only memory. When an instruction is analyzed by the control unit, that instruction will cause a certain sequence of microinstruction words in the ROM to be executed. The result is the generation of logic signals that will carry out the operation designated by the instruction. The instruction set for any digital computer is defined by the operation of the control unit.

The exact logic circuitry used in the control unit varies widely from one machine to another. However, the basic elements are shown in Figure 6-3. The control unit consists of an instruction register, a program counter, an instruction decoder, a clock oscillator, and some type of sequential logic circuit used for generating the control pulses.

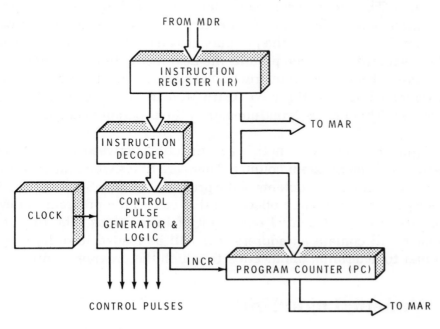

Figure 6-3
Typical control unit organization.

The instruction register is a multi-bit flip-flop register used for storing the instruction word. When an instruction is taken from memory, it passes through the MDR and then into the instruction register. From here, the instruction is decoded by the instruction decoder. This logic circuitry recognizes which instruction is to be performed. It then sends the appropriate logic signals to the control pulse generator. Under the control of the clock oscillator, the control pulse generator then produces the logic signals that will enable the other circuitry in the machine to carry out the specified instruction.

The program counter is simply a binary up counter that keeps track of the sequence of instructions to be executed. The program consists of instructions that are stored in sequential memory locations. To begin a program, the program counter is loaded with the starting address. The starting address is the location of the first instruction in the program to be executed. The first instruction is then read out of memory, interpreted, and carried out. The control circuitry then increments the program counter. The contents of the program counter is then fed to the memory address register that then permits the next instruction in sequence to be addressed. Each time an instruction is executed the program counter is incremented so that the next instruction in sequence is fetched and executed. This process continues until the program is complete.

In Figure 6-3, you will notice a connection between the instruction register and the program counter. There are times when the instruction itself will modify the contents of the program counter. Some instructions specify a jump or branch operation that causes the program to deviate from its normal sequential execution of instructions. The instruction register will contain an address that will be loaded into the program counter to determine the location to which the program jumps.

Arithmetic Logic Unit

The arithmetic logic unit (ALU) is that portion of the digital computer that carries out most of the operations specified by the instructions. It performs mathematical operations, logical operations, and decision-making functions. Most arithmetic logic units can perform addition and subtraction. Multiplication and division operations are generally programmed. The ALU can also perform logic operations such as inversion, AND, OR, and exclusive OR. In addition, the ALU can make decisions. It can compare numbers or test for specific quantities such as zero or negative numbers.

The arithmetic logic unit and control unit are very closely related, so much so that it is sometimes difficult to separate them. Because of this, the ALU and control unit together are often referred to as the central processing unit (CPU). Most microprocessors are single chip LSI CPUs.

The arithmetic-logic unit in a digital computer varies widely from one type of machine to another. Figure 6-4 shows the ALU circuitry associated with a very simple, minimum digital computer. The heart of the arithmetic-logic unit is the accumulator register. It is in this register where most of the computer operations take place. Here, the data is manipulated, computations are carried out, and decisions are made.

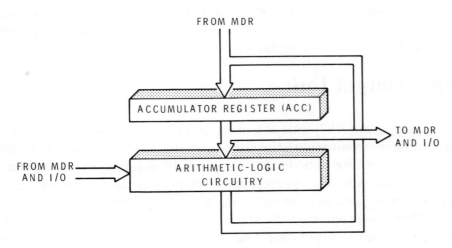

Figure 6-4
Arithmetic logic unit organization.

The accumulator register is a flexible unit that can usually be incremented and decremented. It can also be shifted right or shifted left. Many of the instructions define operations that will be carried out on the data stored in the accumulator register. The size of the accumulator register is generally determined by the basic computer word size, which is the same as the memory word size.

Associated with the accumulator register is the arithmetic logic circuitry. For the most part, this circuitry is a binary adder. With binary adder, both binary addition and subtraction can be accomplished. The arithmetic logic circuitry is also usually capable of carrying out logic operations such as AND, OR, and exclusive OR on the data stored in the accumulator register.

The arithmetic-logic circuitry is capable of adding two binary words. One of the binary words is stored in the accumulator. The other binary word is stored in the memory data register. The sum of these two numbers appears at the output of the arithmetic-logic circuitry and is stored in the accumulator register, replacing the number originally contained there. Most of the other operations with the arithmetic logic circuitry are carried out in this manner. The two words to be manipulated are initally stored in the accumulator and the MDR, with the results of the operation appearing back in the accumulator replacing the original contents.

Input-Output Unit

The input-output (I/O) unit of a computer is that section that interfaces the computer circuitry with the outside world. The term "outside world" refers to everything outside the computer. In order for the computer to communicate with an operator or with peripheral equipment, some means must be provided for entering data into the computer and reading it out. Data and programs to be stored in the memory are usually entered through the input-output unit. The solutions to calculations and control output signals are usually passed to the external equipment through the I/O unit.

The I/O unit is generally under the control of the CPU. Special I/O instructions are used to transfer data into and out of the computer. More sophisticated I/O units can recognize signals from extra peripheral devices called interrupts that can change the operating sequence of the program. Some I/O units permit direct communications between the computer memory and an external peripheral device without interference from the CPU. Such a function is called direct memory access (DMA).

The input/output section of a digital computer is the least clearly defined of all digital computer sections; it can vary from practically no circuitry at all to very complex logic circuitry approaching the magnitude of the remainder of the computer itself. For our explanation of digital computer operation here, we will assume the simplest form of input/output circuitry.

Data transfers between the computer and external peripheral devices take place via the accumulator register. Data to be inputted and stored in memory will be transferred a word at a time into the accumulator and then into the memory through the MDR. Data to be outputted is first transferred from the memory into the MDR, then into the accumulator, and finally to the external peripheral device. These data transfers into and out of the accumulator register take place under the control of the CPU and are referred to as programmed I/O operations. Special input/output instructions cause the proper sequence of operations to take place.

Most digital computers can also perform I/O operations at the request of an **interrupt**, a signal from an external device requesting service. The external device may have data to transmit to the computer or may require the computer to send it data. When an interrupt occurs, the computer completes the execution of its current instruction, and then jumps to another program in memory that services the interrupt. Once the interrupt request has been handled, the computer resumes execution of the main program. Data transfers occurring in the interrupt mode can also take place through the accumulator.

Digital Computer Operation

Now that you are familiar with the basic architecture of a digital computer, you are ready to see how the various sections operate together to execute a program. The units we described previously, when used together, actually form a simple hypothetical digital computer. In this section, it is used to demonstrate how a computer operates. A program for solving a problem is already stored in memory. The computer will execute each instruction until the problem is solved. The complete operation is described and the contents of each register shown as the program is carried out.

Assume that the problem to be solved is a simple mathematical operation that tells us to add two numbers, subtract a third number, store the result, print the answer, and then stop. The numbers that we will work with are 36, 19, and 22. The program calls for adding 36 and 19, subtracting 22, and then storing and printing the answer.

The solution to this simple problem as it is solved step-by-step by the computer is illustrated in Figure 6-5. Here, we show a simplified block diagram of the digital computer, emphasizing the memory and the major registers. The program is stored in memory. The contents of each memory location, either instruction or data, is shown adjacent to the memory address. To solve this problem, the computer sequentially executes the instructions. This is done in a two-step operation. First, the instruction is fetched or read out of memory. Second, the instruction is executed. This fetch-execute cycle is repeated until all of the instructions in the program have been executed.

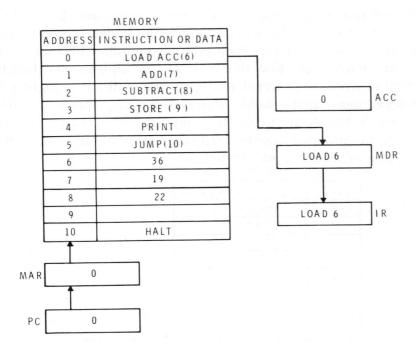

Figure 6-5
Fetch first instruction (LOAD).

In Figure 6-5, the first instruction of the program is fetched. The instruction word is read out of memory and appears in the memory data register (MDR). It is then transferred to the instruction register (IR) where it is interpreted. Note that the memory address register (MAR) contains 0, which is the address of the first instruction. The accumulator register (ACC) is set to 0 prior to the execution of the program.

Figure 6-6 shows the execution of the first instruction, LOAD ACC (6), which tells us to load the accumulator with the data stored in memory location 6. In executing this instruction, the number 36 is transferred to the accumulator. Note how this is done. The address specified by the instruction word (6) is transferred from the instruction register to the memory address register (MAR). This causes the number 36 stored in that location to be transferred to the MDR and then to the accumulator. During this step, the program counter (PC) is incremented by one so the next instruction in sequence will be fetched.

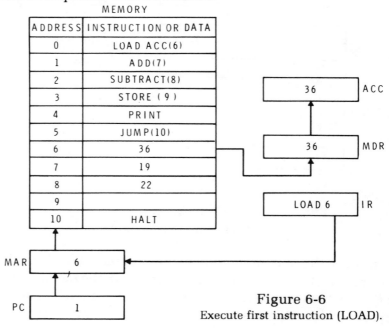

Figure 6-6
Execute first instruction (LOAD).

Figure 6-7 shows the fetch operation for the second instruction. The contents of the program counter is transferred to the MAR so that the ADD(7) instruction is fetched. This instruction passes though the MDR into the instruction register.

The execution of the add instruction is shown in Figure 6-8. This instruction tells us to add the contents of memory location 7 to the contents of the accumulator. The address of the add instruction is transferred to the MAR. This causes the contents of memory location 7, the number 19, to be transferred to the MDR. The contents of the MDR are added to the contents of the accumulator with the sum appearing back in the accumulator. As you can see, the sum of 36 and 19 is 55. Note that the program counter is again incremented so that the next instruction in sequence will be fetched.

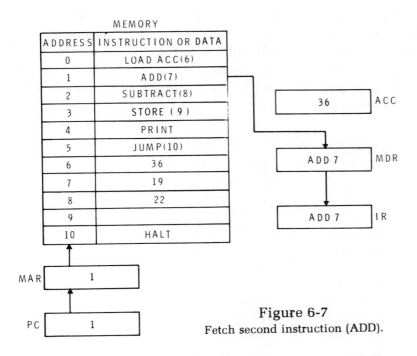

Figure 6-7
Fetch second instruction (ADD).

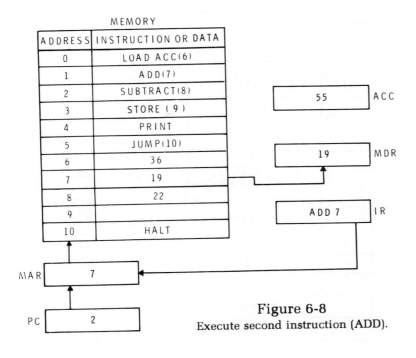

Figure 6-8
Execute second instruction (ADD).

The remaining instructions in the program are fetched and executed in a similar manner. The third instruction, a subtract, causes the memory contents of location 8 to be subtracted from the contents of the accumulator with the resulting remainder appearing in the accumulator. This produces an answer of 33. This fetch-execute sequence is shown in Figures 6-9 and 6-10.

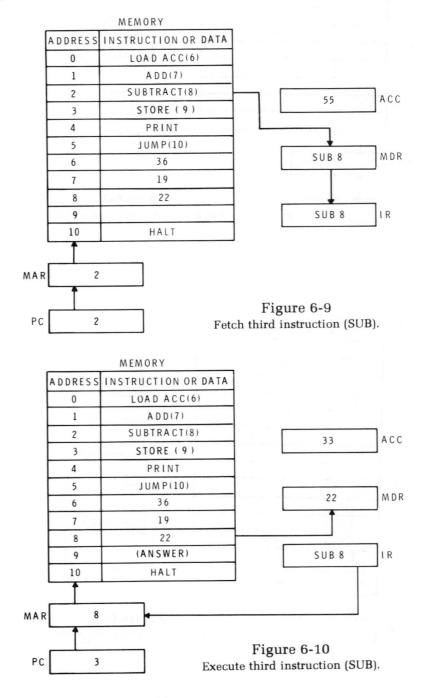

Figure 6-9
Fetch third instruction (SUB).

Figure 6-10
Execute third instruction (SUB).

The next instruction in sequence, STORE(9), tells us to store the contents of the accumulator in memory location 9. The number 33 in the accumulator is transferred to the MDR and stored in location 9, as indicated in Figures 6-11 and 6-12.

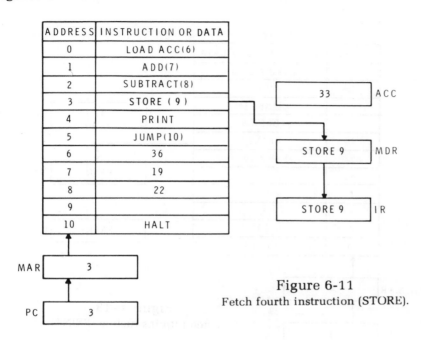

Figure 6-11
Fetch fourth instruction (STORE).

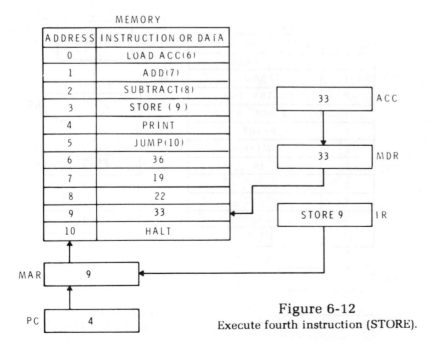

Figure 6-12
Execute fourth instruction (STORE).

The fifth instruction in the program, PRINT, tells us to print the contents of the accumulator on the external printer. The number stored in the accumulator will then be transferred to a printer where it is printed. The fetch-execute cycle for this operation is shown in Figures 6-13 and 6-14.

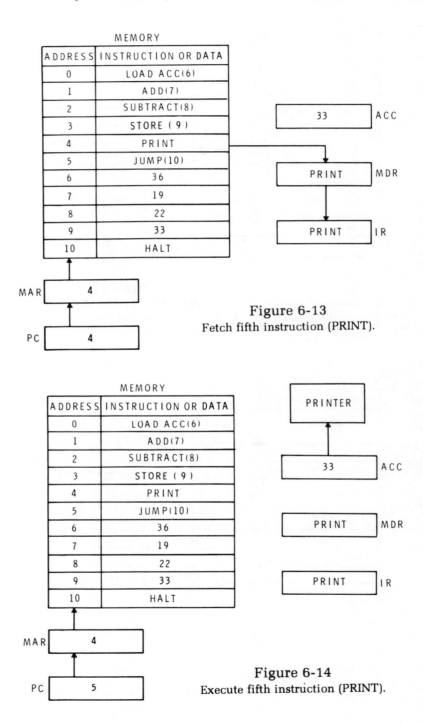

Figure 6-13
Fetch fifth instruction (PRINT).

Figure 6-14
Execute fifth instruction (PRINT).

The sixth instruction in sequence is a JUMP(10) instruction that causes the normal sequence of program executions to change. The jump instruction tells us not to execute the contents of the next memory location in sequence. Instead, it tells us to take the next instruction from memory location 10. You can see by referring to Figure 6-15 that the contents of the next memory location in sequence (address 6) contains a data word.

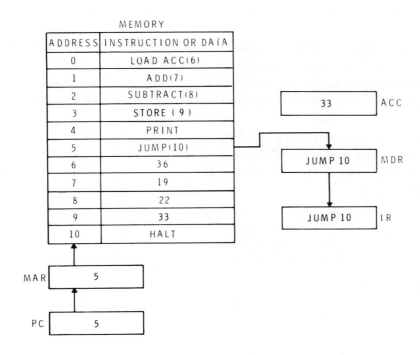

Figure 6-15
Fetch sixth instruction (JUMP).

The computer, being a dumb machine, would simply interpret a data word as an instruction and attempt to execute it. If this ever happens, the result of the computation will be erroneous. The purpose of the jump instruction in our program is to jump over the data words in the program stored in locations 6, 7, 8, and 9. The program is continued in location 10, where a HALT instruction is stored. By executing the jump instruction, the program counter is loaded with the address portion of the jump instruction (10) instead of being incremented as it normally is. See Figure 6-16. This causes the computer to fetch and execute the instruction stored in location 10. This is illustrated in Figures 6-17 and 6-18.

The last instruction in the program is a HALT. This instruction has no effect other than to stop the operation of the machine. Note in Figure 6-18 that the program counter was incremented so that it contains the memory location (11) of the next instruction in sequence to be fetched.

Study the program shown in Figures 6-5 to 6-18. Trace through each of the fetch and execute cycles for each instruction to be sure that you fully understand the operation. All digital computers operate in this same way with minor variations.

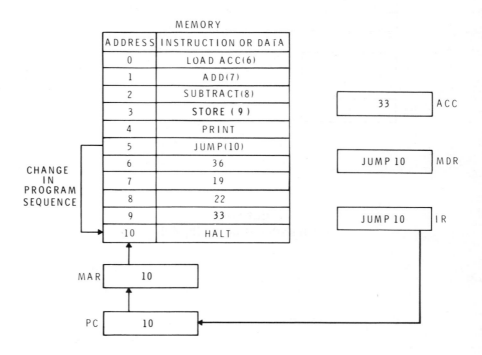

Figure 6-16
Execute sixth instruction (JUMP).

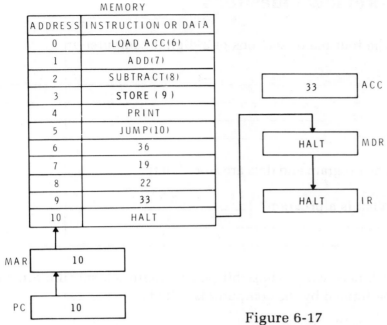

Figure 6-17
Fetch seventh instruction (HALT)

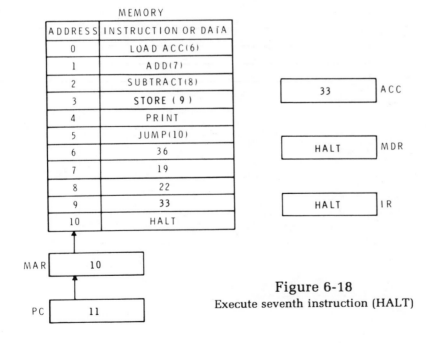

Figure 6-18
Execute seventh instruction (HALT)

Self-Review Questions

8. The four major sections of a digital computer are:

1. _____
2. _____
3. _____
4. _____

9. The program and data are stored in the _____.

10. What is a program? _____
_____
_____

11. A binary word whose bit pattern defines a specific function to be performed by the computer is called a _____.

12. A binary word used to locate a particular word in memory is called an _____.

13. List the four basic sections of the memory.

1. _____
2. _____
3. _____
4. _____

14. What is the control unit of a computer and what is its purpose?

    _____

    _____

    _____

15. What is the arithmetic logic unit?

    _____

    _____

16. The main computational and data manipulation register in a computer is the _____.

17. The arithmetic logic unit and control unit combined are referred to as the _____ _____.

18. What is the input-output unit of a computer? _____

    _____

    _____

19. What is an interrupt? _____

    _____

    _____

20. In carrying out a program, the computer repeats a series of _____ and _____ operations on the instructions in memory.

Self-Review Answers

8. The four major sections of a digital computer are:

   1. Memory
   2. Control Unit.
   3. Arithmetic Logic Unit
   4. Input-Output Unit

9. The program and data are stored in the **memory**.

10. A program is a series of instructions that are stored and executed in sequence to carry out some specific function.

11. A binary word whose bit pattern defines a specific function to be performed by the computer is called a **computer instruction**.

12. A binary word used to locate a particular word in memory is called an **address**.

13. The four basic sections of memory are:

   1. The memory itself
   2. Memory address register (MAR)
   3. Memory address decoder
   4. Memory data register (MDR)

14. The control unit is a sequential logic circuit. Its purpose is to examine each instruction word and generate the control pulses needed to carry out the specific function.

15. The arithmetic logic unit is the portion of the computer that carries out most of the operations specified by the program instructions.

16. The main computational and data manipulation register in a computer is the **accumulator**.

17. The arithmetic logic unit and control unit combined are referred to as the **central processing unit** (CPU).

18. The input-output unit of a computer is the section that interfaces the computer circuitry with the outside world.

19. An interrupt is a signal from an external device requesting service.

20. In carrying out a program, the computer repeats a series of **fetch** and **execute** operations on the instructions in memory.

COMPUTER PROGRAMMING

A digital computer without a program is useless. The logic circuitry making up the computer is incapable of performing any useful end function without a program. It is this characteristic of a digital computer that sets it apart form other types of digital circuitry. And it is this characteristic that makes the digital computer the versatile machine that it is. For this reason, a discussion of digital computers is not complete without information on programming.

The process of using a digital computer is mainly that of programming it. Whether the computer is a simple microprocessor or a large scale system, it must be programmed in order for it to perform some useful service. The application of the computer will define the program. Programming is the process of telling the computer specifically what it must do to satisfy our application.

Programming is a complex and sophisticated art. In many ways, it is almost a field apart from the digital circuitry and the computer hardware itself. There are many different levels of programming and many unique methods that are employed. For that reason, it is impossible to cover them all here. The purpose of this section is to give you an overview of the process of programming a computer. The emphasis will be on programming small scale digital computers such as the microprocessor.

Programming Procedure

There are many different ways to program a digital computer. The simplest and most basic form of programming is **machine language programming**. This is the process of writing programs by using the instruction set of the computer and entering the programs in binary form, one instruction at a time. Programming at this level is difficult, time consuming, and error prone. It also requires an in-depth understanding of the computer organization and operation. Dispite these disadvantages, however, this method of programming is often used for short simple programs. While most computer applications do not use machine language programming, it is desirable to learn programming at this level. It helps to develop a through knowledge of machine operation and generally results in the shortest, most efficient programs. Many microprocessors and minicomputers are programmed in machine language.

To illustrate the concepts of programming in this unit, we will use machine language programming. Other more sophisticated methods of programming will be discussed later.

Programming a digital computer is basically a seven step process. These seven steps are: (1) define the problem; (2) develop a workable solution; (3) flow chart the problem; (4) code the program; (5) enter the program into the computer; (6) debug the program; (7) run the program. We will discuss each of these steps in detail.

The first and perhaps the most important step in programming a digital computer is defining the problem to be solved. The success of the program is directly related to how well you define the operation to be performed. There is no set standard for the problem defining procedure, and you can use any suitable method. The definition can be a written statement of the function to be carried out, or it may take the form of a mathematical equation. In some cases, the problem may be more easily defined by graphical means. For control applications, the problem may be expressed with a truth table. The form in which you place the definition is strictly a function of the application.

Once the problem is analyzed and defined, you can begin thinking of how the computer may solve the problem. Remember that a computer program is a step-by-step sequence of instructions that will lead to the correct results. You should think in terms of solving your problem in some step-by-step sequential manner. What you will be doing in this phase of the programming procedure is developing a algorithm. An **algorithm** is a method or procedure for solving a problem.

An important point to remember is that there is usually more than one way to solve a given problem. In other words, there is more than one algorithm suitable for achieving the goal that you have set. Much of the job of programming is in determining the alternatives and weighing them to select the best suitable approach. The simplest and most direct algorithms are usually the best.

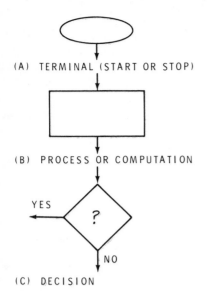

Figure 6-19
Basic flowcharting symbols.

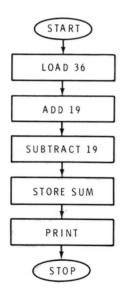

Figure 6-20
Flow chart for the problem 36 + 19 = 33
and print.

The next step is to flowchart the problem. A **flow chart** is a graphical description of the problem solution. Various symbols are used to designate key steps in the solution of the problem. Figure 6-19 shows the basic flowcharting symbols. An oval defines the starting and finishing points. A rectangular box defines each individual computational step leading to the solution. Each rectangle contains some basic operation or calculation that is to take place. The diamond shaped symbol represents a decision point. It is often necessary to observe the intermediate results in a problem solution and make a decision regarding the next step to be taken. There are generally two exits to the diamond-shaped, decision-making symbol. These represent a yes or no type of decision.

Figure 6-20 shows a simple flow chart for the problem solved earlier. No decision was made in this program.

As you can see, the flow chart is a graphical representation of the basic method used to solve the problem. The flow chart permits you to visualize the algorithm you developed. In many cases, the flowcharting of a problem helps to determine the best approach to solve a problem since it forces you to think in a logical sequence and express the solution in a step-by-step form.

At this point in the programming procedure, your problem is quite well defined and a basic method of solving the problem has been determined. You are now ready to convert your flow chart and algorithm into a machine language program. This process is called coding. **Coding** is the procedure of listing the specific computer instruction sequentially to carry out the algorithm defined by the flow chart. This requires a familiarity with the instruction set of the computer you plan to use.

The next step in the programming procedure is to load the program into the computer memory. Once you have written the program with the computer instructions, you have all of the information necessary to enter that program into the computer memory. If you are dealing with machine language programming, you will convert the instruction words into their binary equivalents and then load them into the computer. If the program is a simple one, it can be loaded by using the binary switches on the front panel of the computer. However, for long, complex programs, this manual loading procedure is difficult and time consuming.

Most computers make it easy for the programmer to enter his program. Because of the availability and use of support programs residing within the machine, the program can usually be entered automatically. One of the most common ways of entering data into a comupter is with a keypunch machine. This is a typewriter-like machine that punches a standard computer card with the instructions to be entered. Teletype input/output machines using perforated paper tape are also commonly used for program entry. The instruction designations are typed on the machine and as they are typed, a paper tape is punched.

Once the cards or paper tape are punched, they are then fed into a tape reader or card reader and loaded into the computer memory. A special program residing within the computer memory, called a loader, causes the program to be loaded automatically.

With the program now in the computer memory, you can begin to run it. However, before you use it to obtain your final answer, it is often necessary to run through the program slowly a step at a time to look for programming errors and other problems. This process is called debugging. You test the program to see that it produces the desired results. If the program produces the correct result, it is ready to use. Often, programming mistakes are encountered and it is necessary to modify the program by changing the instruction steps. Often the entire program may be discarded and a new one written, using a different algorithm.

Once the program has been debugged, it is ready for use. With the program stored in memory, your problem can be solved. The computer is started and the desired results are produced.

Writing Programs

Before you can begin coding programs, you must be familiar with the instruction set of the computer you are using. Most digital computer instruction sets are basically alike in that they all perform certain basic functions such as addition, branching, input/output and the like. But each instruction set is different because the logic circuits unique to each computer carry out these operations in different ways. To code the program properly, you must know exactly what each instruction does. You can get this information by studying the instruction set as it is listed and explained in the computer's operation and programming manuals. By studying the instructions set, you will learn how the computer is organized and how it operates. The insight you gain from this will be valuable to you not only in coding the program but also in developing the best solution to a problem with a given machine.

Computer Instructions

A computer instruction is a binary word that is stored in the computer memory and defines a specific operation that the computer is to perform. The instruction word bits indicate the function to be performed and the data which is to be used in that operation.

There are two basic types of computer instructions: memory reference and non-memory reference. A memory reference instruction specifies the location in memory of the data word to be used in the operation indicated by the instruction. A non-memory reference instruction simply designates an operation to be performed. Non-memory reference instructions generally refer to internal housekeeping operation to be performed by the computer and manipulations on data stored in the various registers in the computer.

Figure 6-21 shows typical instruction word formats for an 8-bit microprocessor. The format shown in Figure 6-21A is a memory reference instruction. The instruction is defined by three 8-bit words which are stored in sequential memory locations. The first 8-bit word is the op code (or operations code) which is simply a binary bit pattern specifying some operation. The second and third 8-bit words specify the memory address of the data or operand to be used. The 8-bit op code defines 256 possible operations or functions. It is the op code that designates the operation that is to be performed. The 16-bit address specifies the memory location of the data to be operated upon. The size of the address generally indicates the maximum memory size of the computer. With 16 bits of address information, $2^{16} = 65,536$ words can be directly addressed. We usually say that the maximum memory size is 65K.

The word format in Figure 6-21B is the typical format for non-memory reference instructions. Only a 8-bit op code is used. In this type of instruction, an address is not needed since we do not reference a location in memory where data is stored. Instead, the bits in this field are used to specify various operations that are to take place within the CPU. For example, such an instruction might call for the resetting (clearing) of a register or the transfer of data from one register to another. Certain types of input-output instructions have this format.

Another instruction type is the immediate instruction which is widely used in microprocessors. The format for this instruction is shown in Figure 6-21C. It consists of an 8-bit op code that specifies the operation. The second 8-bit part of this instruction is the data or operand to be used in the operation called for. The immediate instruction is like the memory reference instruction in that it specifies the use of some data word. The data to be used is in the instruction word itself rather than being referenced by an address in the instruction word. Immediate instructions save memory space and shorten instruction fetch and execution times.

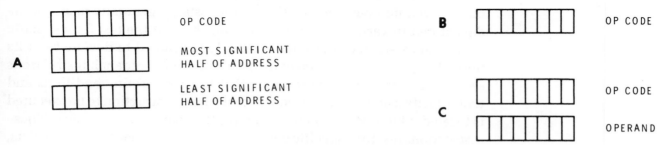

Figure 6-21

Typical computer or microprocessor instruction formats.

(A) memory reference,

(B) nonmemory reference,

(C) immediate.

Another method of classifying computer instruction is to group them according to the type of functions that they perform. These include arithmetic and logic, decision-making, data moving, and control. We will consider each of these in more detail.

An arithmetic instruction defines a specific mathematical operation that is to take place. The most commonly used arithmetic instructions are add and subtract. In larger computers, the multiply and divide functions are also included. Multiply and divide operation in smaller computers such as minicomputers and microprocessors are carried out by special subroutines. As an example, multiplication can be performed by repeated addition. Division can be programmed by the use of repeated subtractions. Arithmetic instructions are generally of the memory reference type.

Logical instructions specify digital logic operations that are to be performed on computer data. These include the standard logic functions of AND, OR and invert(complement). Many computers include the exclusive OR function. Other logic instructions include shift right and shift left operations. The AND, OR and XOR logical instructions are usually memory reference type. The shifting and inversion instructions are of the non-memory reference type, as they generally refer to operations carried out on data stored in one of the computer's registers.

A decision-making instruction is one that permits the computer to test for a variety of results and, based upon these tests, make a decision regarding the next operation to be performed. It is the decision-making instructions that set the computer apart from the standard calculator and allow the computer to automate its operations. A decision-making instruction generally follows a sequence of other instructions that perform some arithmetic or logical operation. Once the operation is performed, the decision-making instruction tests for specific results. For example, decision-making instructions test for positive or negative numbers, zero, odd or even numbers, or equality. These tests are generally made on the data stored in various registers in the machine. If the test for a specific condition exists, the computer is usually instructed to deviate from its normal sequential execution of instructions. Jump or branch instructions are memory reference instructions that test for certain conditions and then specify a memory location where the next instruction to be executed is located. Skip instructions also change the computing sequence. These instructions test for a specific condition and then, if that condition exists, direct the computer to skip the next instruction in sequence. Skip instructions are non-memory reference types.

A data moving instruction is one that causes data words to be transferred from one location to another in the computer. It is these instructions that are used to take data from memory and load it into one of the operating registers in the computer. Other data moving instructions cause data stored in a register to be stored in a specific memory location. These are memory reference instructions. Other data moving instructions specify the transfer of data words between registers in the machine. These are non-memory reference instructions. The data moving instructions provide a flexible means of transferring data within the machine to prepare it to be processed as required by the application.

A special class of data moving instructions are the input/output instructions. I/O instructions cause data to be transferred into and out of the computer. These non-memory reference instructions often specify one of several input/output channels or a specific peripheral device. Input/output operations can be programmed to take place through the operating registers of the machine, or in some computers, directly between the memory and the peripheral unit.

A control instruction is a non-memory reference instruction that does not involve the use of data. Instead, it designates some operation that is to take place on the circuitry in the computer. Clearing a register, setting or resetting a flip-flop or halting the computer are examples of control instructions.

A Hypothetical Instruction Set

A typical but hypothetical instruction set for a minicomputer or microprocessor is shown in Table I. Only a few of the most commonly used instructions are listed so that you can become acquainted with them quickly. Real instruction sets are far more extensive. Nevertheless, the instruction set in Table I is representative. We will use it to demonstrate the writing and coding of programs.

The instruction set in Table I can apply to the hypothetical computer described earlier or a typical microprocessor. For the instructions listed here, we assume that the computer has an 8-bit word length and 65K of memory. The accumulator and memory data registers are 8 bits in length. The program counter and MAR are 16 bits in length. I/O transfers take place though the accumulator. A single instruction may occupy one, two or three consecutive memory locations depending upon its format as shown in Figure 6-21. Study the instructions in Table I so that you will be familiar with the operation each performs. Note that each instruction is designated by a three-letter mnemonic. The type of instruction is designated by the letters R (memory reference), N (non-memory reference), A (arithmetic-logic), T (data moving or transfer), D (decision) and C (control). Despite the simplicity of this instruction set, it can be used to program virtually any function.

HYPOTHETICAL COMPUTER INSTRUCTION SET

Table I

MNEMONIC	TYPE OF INSTRUCTION	OPERATION PERFORMED
LDA	R, T	Load the data stored in the specified memory location (M) into the accumulator register.
STA	R, T	Store the data in the accumulator register in the specified memory location (M).
ADD	R, A	Add the contents of the specified memory location (M) to the contents of the accumulator and store the sum in the accumulator.
SUB	R, A	Subtract the contents of the specified memory location (M) from the contents of the accumulator and store the remainder in the accumulator.
AND	R, A	Perform a logical AND on the data in the specified memory location (M) and the contents of the accumulator and store the results in the accumulator.
OR	R, A	Perform a logical OR on the data in the specified memory location (M) and the contents of the accumulator and store the results in the accumulator.
JMP	R, D	Unconditionally jump or branch to the specified memory location (M) and execute the instruction stored in that location.

JMZ	R, D	Jump to the specified memory location if the content of the accumulator is zero (reset). Execute the instruction stored in that location. If the accumulator is not zero, continue with the next instruction in normal sequence.
CLA	N, C	Clear or reset the accumulator to zero.
CMP	N, A	Complement the contents of the accumulator.
SHL	N, A	Shift the contents of the accumulator one bit position to the left.
SHR	N, A	Shift the contents of the accumulator one bit position to the right.
INP	N, T	Transfer an 8-bit parallel input word into the accumulator.
OUT	N, T	Transfer the contents of the accumulator to an external device.
HLT	N, C	Halt. Stop computing.
INC	N, C	Increment the contents of the accumulator.
DCR	N, C	Decrement the contents of the accumulator.
SKO	N, D	If the number in the accumulator is odd (LSB = 1), skip the next instruction and execute the following instruction. If the accumulator is even (LSB = 0), simply execute the next instruction in sequence.

Example Programs

The following examples illustrate the use of the instruction set in writing programs. The program description, flowchart and instruction code are given in each example. Study each program, mentally executing the instructions and imagining the outcome. The format of the instruction coding is shown below.

3 ADD (7)

The number on the left is the memory address. The mnemonic specifies the instruction. The number in parenthesis is the address of the operand called for by a memory reference instruction. This line of instruction coding says that memory location 3 contains an add instruction that tells us to add the content of location 7 to the content of the accumulator.

The program shown below is a repeat of the program given in Figures 6-5 to 6-18. The only differences are the memory location numbers of the instructions, the use of mnemonics, and the substitution of the OUT instruction for the PRINT instruction.

```
 0   LDA   (16)
 3   ADD   (17)
 6   SUB   (18)
 9   STA   (19)
12   OUT
13   JMP   (20)
16   36
17   19
18   22
19   ANSWER
20   HLT
```

The difference in memory addresses is the result of the assumption that our computer uses an 8-bit word and that memory reference instructions occupy three sequential memory locations. In the program of Figures 6-5 to 6-18, we assumed that one memory address contained one instruction. In the program above, the LDA (16) occupies memory locations 0, 1 and 2. The op code is in 0, the most significant part of the address (0000 0000) is in location 1, and the least significant part of the address (0001 0000) is in location 2. The ADD, SUB, STA and JMP memory reference instructions each occupy three sequential locations. The OUT and HLT instructions do not reference memory so they occupy only a single location.

The program below illustrates the use of the logical instructions.

This program is designed to implement the logical NOR function. Since the computer instruction set contains only the AND and OR instructions, it is necessary to complement the OR function. Figure 6-22 shows the flow chart, and the program is shown below. The algorithm is $\overline{A + B}$.

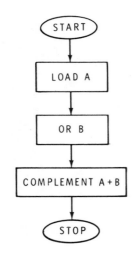

Figure 6-22
Flow chart for $\overline{A + B}$.

0	LDA	(11)
3	OR	(12)
6	CMP	
7	STA	(13)
10	HLT	
11	A	
12	B	
13	ANSWER	

The next program illustrates several important concepts. First, it shows how the computer makes decisions. Second, it demonstrates the use of a program loop. A **loop** is a sequence of instructions that is automatically repeated. The sequence is executed once and a jump instruction causes the program to branch back (loop) to the beginning of the sequence and repeat it again.

The program below is designed to enter two 4-bit BCD numbers and store them in a single 8-bit memory location. The desired memory format is shown in Figure 6-23. The BCD digits are entered, one at a time, into the

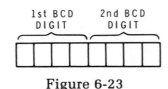

Figure 6-23
Memory format for two BCD digits.

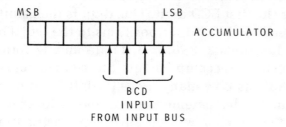

Figure 6-24
Loading the accumulator from the data
bus.

four least significant bit positions of the accumulator as shown in Figure 6-24. The first digit entered must be moved to the four most significant bit positions of the accumulator before the second digit can be entered. This is accomplished with a series of shift left instructions.

The flow chart in Figure 6-25 shows how this is accomplished. The program is given below.

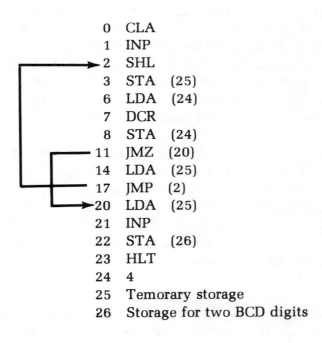

0	CLA	
1	INP	
2	SHL	
3	STA	(25)
6	LDA	(24)
7	DCR	
8	STA	(24)
11	JMZ	(20)
14	LDA	(25)
17	JMP	(2)
20	LDA	(25)
21	INP	
22	STA	(26)
23	HLT	
24	4	
25	Temorary storage	
26	Storage for two BCD digits	

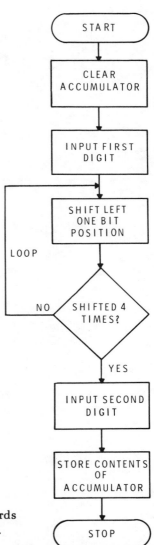

Figure 6-25
Flow chart showing how two BCD words
are stored in one memory location.

The first instruction (CLA) clears the accumulator. The second instruction (INP) loads the first BCD digit. This digit is then shifted left one bit position. We need to shift it four positions to the left. The sequence of instructions in locations 2, 3, 6, 7, 8, 11, 14 and 17 form a loop and a decision-making test to accomplish this. Stored in memory location 24 is a number that tells us how many times to shift. That number is loaded into the accumulator decremented by one and restored each time a shift occurs. We test that number with a jump on zero instruction (JMZ). When the BCD digit has been shifted four times, the number in location 24 has been reduced to zero. The JMZ instruction detects this condition and branches the program to location 20 where the data word is retrieved from temporary storage and the second BCD digit is inputted.

Now, we will consider the loop and decision-making process in more detail. After the first input digit is loaded, it is shifted left once. We then store it temporarily in location 25. This is to prevent loss of the data while we are in our decision-making loop. Next, the number of desired shifts is loaded into the accumulator. We decrement it by one, indicating that we have shifted left once. Next, we restore this number (now 3) in location 24. This number, which is still in the accumulator, is now tested with the JMZ instruction. At this time the accumulator is not zero so the program does not branch. Instead, the next instruction in sequence is executed. This is a load accumulator instruction that retrieves the data words which we temporarily stored in location 25. Then the JMP instruction is executed. This instruction returns us to location 2 to produce another shift. It is the jump instruction that creates the loop.

The loop is then repeated three more times. On the fourth pass through the loop, the number in location 24 is decremented to zero. The JMZ instruction detects this condition and causes the program to branch to location 20. We have now escaped from the loop. Next we reload the shifted data from location 25. Finally, we load the second BCD digit. Both digits are now in the accumulator so we can store them in location 26 with the STA (26) instruction. The program is now complete and the HLT instruction terminates it.

These examples show how a computer performs its work. It does it laboriously, one step at a time. The only thing that makes it practical is its high speed operation. With each instruction taking only microseconds, even long complex programs are executed quickly. To an operator, the execution appears almost instantaneous.

Self-Review Questions

21. What is machine language programming? _____
_____
_____

22. List the seven steps of programming.

    1. _____
    2. _____
    3. _____
    4. _____
    5. _____
    6. _____
    7. _____

23. A procedure for solving a problem is an _____
_____

24. A graphical description of the problem solution is known as a
_____ _____.

25. What is coding? _____
_____
_____

26. Name the two types of computer instructions and describe each
one. _____
_____
_____
_____
_____
_____
_____

27. Name the computer instruction groups according to function.

    1. _____
    2. _____
    3. _____
    4. _____
    5. _____

28. What is a loop? _____

_____

_____

29. The content of the accumulator is 10111010. The CMP instruction is executed. The new accumulator content is _____.

30. Which instruction would you use to down-count the accumulator? _____.
(Use the Hypothetical Computer Instruction Set.)

31. The content of the accumulator is 45. A JMZ (34) instruction in location 25 is then executed. The next instruction executed is in location _____.

32. Program loops are implemented with the _____ and _____ instructions.

33. The number 0110 0101 is stored in the accumulator. The number in memory location 18 is 1111 0000. The AND (18) instruction is executed. The content of the accumulator becomes _____.

34. Study the program below. At the completion of the program, the content of the accumulator is _____.

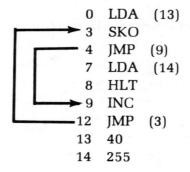

```
 0   LDA   (13)
 3   SKO
 4   JMP   (9)
 7   LDA   (14)
 8   HLT
 9   INC
12   JMP   (3)
13   40
14   255
```

Self-Review Answers

21. Machine language programming is the process of writing programs by using the instruction set of the computer and entering the programs in binary form, one instruction at a time.

22. The seven steps of programming are:

    1. Define the program.
    2. Develop a solution.
    3. Flow chart the problem.
    4. Code the program.
    5. Enter the program into the computer.
    6. Debug the program.
    7. Run the program.

23. A procedure for solving a problem is an **algorithm**.

24. A graphical description of the problem solution is known as a **flow chart**.

25. Coding is the procedure of listing the specific computer instructions sequentially to carry out the algorithm defined by the flow chart.

26. The two types of computer instructions are memory reference and nonmemory reference. A memory reference instruction specifies the location in memory of the data word to be used in the operation indicated by the instruction. A nonmemory reference instruction simply designates an operation to be performed.

27. The computer instruction groups according to function are:

    1. Arithmetic
    2. Logic
    3. Decision making
    4. Data moving
    5. Control

28. A **loop** is a sequence of instructions that is automatically repeated.

29. The complemented accumulator content is 01000101.

30. To down-count the accumulator, you should use DCR (decrement).

31. The next instruction executed is in location 28.

   The JMZ(34) instruction tests for a zero accumulator. The accumulator content is 45; therefore, the program does not branch to location 34. Instead, it executes the next instruction in sequence, which begins in location 28. Remember that the JMZ(34) instruction and its reference address occupies locations 25, 26 and 27.

32. Program loops are implemented with the JMP and JMZ instructions.

33. The content of the accumulator becomes 0110 0000. Consider the two words to be ANDed as inputs to an AND gate as they would appear in a truth table. Then AND each corresponding pair of bits to get the result.

34. The content of the accumulator is 255.

   This program uses the "skip on odd" accumulator instruction (SKO) to test the content of the accumulator. It first loads the content of location 13 into the accumulator. This is the number 40. The SKO then tests for an odd condition by monitoring the LSB. Since 40 is even, the program does not skip. The next instruction in sequence is executed. This is a JMP (9) instruction which causes the program to branch to location 9. Here the INC instruction is executed. The accumulator then becomes 41. The next instruction JMP (3) loops the program back to location 3, where the SKO instruction again tests the accumulator. This time, the content is odd. The instruction in location 4 is now skipped and the next in sequence is executed. This is an LDA (14) which loads 255 into the accumulator. The program then halts.

SOFTWARE

The steps that we have just described make up the procedure typically used in developing application programs for the digital computer. The program may be solving a mathematical equation, sorting and editing a large volume of data, or providing some type of automatic control to an external machine.

These application programs fall into a larger category of computer programs called software. **Software** is a general term used to describe all of the programs used in a digital computer. Besides the specific applications programs, there are many special programs supplied by the computer manufacturer which are used to simplify and speed up the use of the computer. These support programs eliminate much of the drudgery from programming and using a computer. It was determined early in the development of the digital computer that the computer itself with special internal control programs could assume much of the responsibility for the detailed translation of a problem into the binary language of the computer.

Subroutines

Many digital computer manufacturers supply what are called software libraries of subroutines and utility programs. A **subroutine** is a short machine language program that solves a specific problem or carries out some often-used operation. For example, typical subroutines in many minicomputers and microprocessors are the multiply and divide programs. Instead of using the multiply subroutine each time it is required in a problem, the subroutine is stored in the computer memory only once. This saves a substantial amount of memory space. Each time the multiply subroutine is required, a jump instruction in the program causes the program to branch to the multiply subroutine. Once the multiplication has been performed, the computer jumps back to the normal program sequence.

There are many different types of commonly used subroutines. Multiplication and division are two of the most commonly used. Other subroutines include binary to BCD and BCD to binary code conversions. To communicate with external peripheral devices which use the decimal number system and the alphabet, a code such as ASCII is used. Data is entered into the computer in the ASCII code. In order for the computer to process this data, it must first be converted into pure binary numbers. Solutions to computer programs are in the pure binary form. A subroutine is used to convert the binary numbers into the ASCII format and then they are sent to an external peripheral device such as the printer.

Utility Programs

Utility programs refer to the short routines used to run the computer. Input/output programs for specific types of peripheral devices fall into this category. A loader is another utility program. This is a short sequence of instructions that allows data to be loaded into the computer. In order to operate, a computer must be programmed. But to load a program into the computer automatically requires that the loader program exist in the memory to begin with. Such loader programs are often entered manually from the computer front panel. The short loader program in memory then permits longer, more complicated programs to be loaded automatically.

Assembler

The most sophisticated software supplied with most computers are large complex conversion programs called assemblers and compilers. These programs allow the computer to be programmed in a simpler language. Machine language programming is completely impractical for many modern applications. To simplify computer programming and eliminate the need for a knowledge of binary numbers and the computer architecture, computer manufacturers have developed easier methods of programming the computer. These methods involve higher level languages, which are special systems for speeding up the programming process. The higher level language permits someone with no computer expertise whatsoever to use the computer. The higher level language permits the programmer to express his problem as a mathematical equation or in some cases as an English language statement. These equations and statements are then fed to the computer, which then automatically converts them into the binary instructions used to solve the problem.

The simplest form of higher level programming language is called assembly language. This is a method of programming the computer an instruction at a time as you do in machine language programming. However, instead of binary designations for each instruction, short multiletter names called mnemonics are given to each of the computer instructions. These are then written sequentially to form the program. Mnemonics are also given to memory addresses to avoid the use of specific memory locations.

Once the computer program is written in the assembly language, it is then entered into the machine along with an assembler program. The assembler program resides in the computer memory and is used to convert the mnemonics into the binary instruction words that the computer can interpret. As you can see, the assembler is a program that eliminates the necessity of dealing with binary numbers in the digital computer. However, since the machine is still programmed an instruction at a time, it provides wide flexibility in solving a given problem.

Compiler

The compiler, like the assembler, is a complex conversion program that resides in the computer memory. Its purpose is to convert a simplified statement of the program into the binary machine code that the computer can understand. The difference between the compiler and the assembler is that the compiler is capable of recognizing even simpler problem statements.

In one type of compiler programming language know as Fortran, the program can be written as an algebraic equation. The algebraic equation is then entered into the computer through a teletypewriter or via punched cards. The compiler program then analyzes the formula and proceeds to construct a binary program to solve this equation at some location in memory.

Another higher level programming language know as Cobol uses English language statements to describe the problem. These English language statements are punched into cards and then read into the computer memory. The compiler interprets them and converts them into the binary program. Unlike an assembly language program which has a one-to-one correspondence of instruction steps with machine language, a single compiler language program statement often causes many binary instructions to be generated.

There are many different types of higher level programming language used with computers. All of them have the prime function of simplifying the programming procedure. They greatly speed up and expedite the communications with the computer. They allow anyone who is capable of defining his problem to use the computer.

Cross Assemblers and Compilers

There are several special types of higher level programming language that have been developed to aid in programming microprocessors. For simple applications, microprocessors are programmed at the machine language level. However, when longer or more complex programs are required, it is desirable to use an assembler or compiler if it is available. For minicomputers and larger scale computers, compilers and assemblers that reside within the computer memory itself are available to aid in the programming process. However, most microprocessors do not have sufficient memory to accomodate such large complex programs. In addition, the microprocessor is generally to be dedicated to a specific application and, therefore, its memory will only be large enough to hold the application program required. In order to simplify the development of programs for use in a microprocessor, special programs called cross-assemblers and cross-compilers have been developed. These are special programs that reside in the memory of a larger, general purpose digital computer. The application programs are written in these higher level languages and the larger machine then converts the application program into the binary machine language required by the microprocessor. The output of the larger scale computer is generally a paper tape containing the binary program, which is later loaded into the microprocessor memory.

Some of the larger more sophisticated microprocessors have been used as the primary component in a microcomputer that can be used as a software development system for that microprocessor. A large random access memory is added to the microprocessor along with appropriate peripheral devices. Resident assembler programs have been developed for these machines. In this way, the microcomputer based on the microprocessor can be used to develop application programs that will be used later in another system employing the same microprocessor.

Self Review Questions

35. What is software? _____
_____
_____

36. What is a subroutine? _____
_____
_____

37. Input/output and loader programs are called. _____
programs.

38. The program used to convert an instruction by instruction mnemonic program into binary machine language is called an _____.

39. What is a compiler? _____
_____
_____

40. What are cross assemblers and compilers? _____
_____
_____

Self-Review Answers

35. Software is a broad term used to describe all of the programs used in a digital computer.

36. A subroutine is a short, machine language program that solves a specific problem or carries out some often used operation.

37. Input/output and loader programs are called **utility** programs.

38. The program used to convert an instruction-by-instruction mnemonic program into binary machine language is called an **assembler**.

39. A compiler, like an assembler, converts a statement of the program into a binary machine language program. However, a compiler is more complex in that it may convert a single statement into several machine language steps. Whereas, the assembler can only convert each mnemonic instruction into a single machine language step.

40. Cross assemblers and compilers are used with large scale, general purpose computers to develop machine language programs for microprocessors.

MICROPROCESSORS

As indicated earlier, a microprocessor is the simplest and least expensive form of digital computer available. However, this section is more specific. It discusses exactly what a microprocessor is, the types that are available, and how they are used.

Types of Microprocessors

Most microprocessors are the central processing unit (CPU) of a digital computer. That is, microprocessors usually contain the arithmetic-logic and control sections of a small scale digital computer. Most of these microprocessors also contain a limited form of input-output circuitry which permits them to communicate with external equipment. To make the microprocessor a complete computer, external memory and input-output devices must be added. An external read only memory (ROM) is normally used to store the program to be executed. Some read/write, random access memory (RAM) may also be used. The external input-output circuitry generally consists of registers and control gating that buffer the flow of data into and out of the CPU.

Microprocessors come in a wide variety of forms. However, the most popular and widely used microprocessor is a MOS LSI circuit. These circuits are made with both P-channel or N-channel enhancement mode MOS devices. The entire CPU is contained on a single chip of silicon and mounted in either a 16, 24 or 40-pin dual in-line package. Such microprocessors are available with standard word lengths of 4, 8 or 16 bits. Other more sophisticated types of microprocessors are contained within two or more integrated circuit packages. When combined, they form a complete, small scale digital computer.

While most microprocessors are of the single chip MOS variety, there are numerous bipolar microprocessors available. These are inherently faster than the MOS devices but occupy more chip space and consume more power. Where high speed is required, these bipolar devices can be used. A recently developed integrated circuit technology, referred to as integrated injection logic (I^2L), combines both the speed of bipolar devices and the high density characteristics of MOS devices. These new I^2L LSI circuits offer many benefits, and their potential for microprocessor applications is great.

Microprocessors can also be constructed with standard TTL integrated circuits. Standard MSI packages can be combined to construct a small CPU. Figure 6-26 shows a computer of this type. While this kind of microprocessor takes more circuitry and consumes more power, it generally offers several advantages. First, the microprocessor can be constructed to execute a special instruction set designed specifically for the application. With a standard off-the-shelf CPU, the instruction set is fixed. Special instruction sets are often necessary for some applications and they can be readily optimized with a special TTL microprocessor design. Another advantage of an MSI TTL microprocessor is high speed. A standard MOS microprocessor may be too slow for the application. The fastest available MOS microprocessor can execute a single instruction in approximately 2 microseconds. The simpler and less sophisticated MOS microprocessors have instruction execution speeds in the 10 to 50 microsecond region. With a special TTL MSI microprocessor, execution speeds in the nanosecond region are easily obtained.

Figure 6-26

A microprocessor made with TTL MSI and SSI integrated circuits. This machine is more powerful than the typical LSI microprocessor but less powerful than a full minicomputer. (Photo courtesy Computer Automation Inc.)

In order to use a standard single chip microprocessor, some form of external memory must be used. The program to be executed by the microprocessor is generally stored in a ROM. Data is stored in RAM. Other external components needed to support a microprocessor are an external clock circuit, input-output registers, and peripheral devices.

All single chip microprocessors incorporate a data bus through which all external data transfers take place. This may be a 4 or 8-bit bi-directional bus over which all data transfers between the memory and input-output devices communicate with the CPU. A bus design of this type greatly minimizes the number of interconnections required to connect the microprocessor to the external devices. The limiting factor of such interconnections is the number of pins on the integrated circuit package. The bus organizations keeps the pin count to a minimum, but at the same time requires time sharing of the bus. Since all data transfers between the memory and CPU and between the CPU and the peripheral devices must use the same input-output lines, each operation must take place at a different time.

The input-output devices used with most microprocessors are quite different from those used with larger digital computers. Most larger computers are connected to input-output devices like CRT terminals, teletypewriters, paper tape readers and punches, card readers and line printers. On the other hand, microprocessors are interfaced to devices such as keyboards, 7-segment LED displays, thumbwheel switches, relays, analog-to-digital and digital-to-analog converters, temperature sensors, and other such components.

Applications of Microprocessors

Microprocessors are used primarily for dedicated functions. Rarely are microprocessors used to implement a general purpose digital computer. The program of a microprocessor is usually stored in the read only memory. This means that the program is fixed and dedicated to the specific application.

There are two general applications for modern LSI microprocessors. They can be used as replacements for minicomputers or as replacements for random hard-wired logic. The development of the minicomputer enabled many engineers to design digital computers into special control systems. The minicomputer was dedicated to the control application and its programmable flexibility offered many benefits. But its cost was very high. Some microprocessors have nearly as much computing power and capability as or minicomputer and can replace the minicomputer in many systems. A microprocessor has the advantage of smaller size, lower cost, and lower power consumption.

Another common use for the microprocessor is as an alternative to standard hard-wired digital logic circuits. Equipment customarily constructed with logic gates, flip-flops, counters, and other SSI and MSI circuits can often be implemented with a single microprocessor. All of the standard logic functions such as logic operations, counting and shifting can be readily carried out by the microprocessor through programming. The microprocessor will execute instructions and sort subroutines that perform the same logic functions.

Many benefits result from using the microprocessor in replacing hard-wired random logic systems. Some of these advantages are: (1) reduced development time and cost; (2) reduced manufacturing time and cost; (3) enhanced product capability; (4) improved reliability.

Development time and cost can be significantly reduced when a microprocessor is used. The design procedures used with standard logic circuits are completely eliminated. Much of the breadboarding, cut-and-try and prototype construction is completely eliminated. Design changes can be readily incorporated and new functions implemented by simply changing the program. With a microprocessor, the logic and control functions are implemented with programs. The program can be written and entered into memory and then tested. System changes are easy to make by simply rewriting the program. Unique functions can be readily added by increasing the size of the program. In many cases, the system can be made self-checking by programming special diagnostic routines.

Development time is further reduced because a single integrated circuit microprocessor usually replaces many other integrated circuits. This reduces wiring and interconnections and simplifies printed circuit board layout. Often the printed circuit board will be significantly smaller with a microprocessor system. Power consumption and cooling are also usually simplified. The benefit of reduced development time and cost, of course, is that the product can come to market or be applied sooner.

Manufacturing costs are also reduced as a result of replacing random logic with a microprocessor. Fewer integrated circuits and smaller printed circuit boards are required to construct the system. Therefore, less time and materials are required to assemble the equipment. The programmed nature of the microprocessor system also makes it easier to test and debug than an equivalent hard-wired system.

Enhanced product capability is another benefit of using the microprocessor to replace hard-wired logic. The power of a digital system implemented with a microprocessor is limited strictly by the imagination of the designer. Many unique features and capabilities can be incorporated into the design by simply adding to the program. The incremental cost for adding such features to a microprocessor system is small compared to that of a hard-wired logic system. The ROM used to store the program usually contains extra room for program additions. Therefore, it is very easy to add special features. Many of these special or unique features would be difficult to incorporate in a random hard-wired logic design because of the extra design time, the complexity, and the additional cost. When a microprocessor is used, no additional parts or significant amount of design time are required to add them. The more unique and special features that a product can incorporate the better it performs and the more competitive it will be in the marketplace.

Another benefit of using the microprocessor to implement digital systems is increased reliability. Whenever the number of integrated circuits and wiring interconnections are reduced in a system, reliability increases significantly. Most system failures result from the failure of an integrated circuit or from an interconnection. The number of integrated circuits and interconnections are greatly reduced in going from a standard hard-wired logic system to a microprocessor system. Increased reliability means fewer failures and leads to a corresponding reduction in both warranty and service costs.

The benefits of using a microprocessor are so significant that they will soon replace most random hard-wired logic designs. But the biggest present disadvantage of using a microprocessor is the designer's lack of programming knowledge. Very little circuit or logic design is required to implement a system with a microprocessor. Instead, the primary skill required is digital computer programming. Most engineers and digital designers were not trained in this subject, and therefore, initial design attempts with microprocessors may be slow and frustrating. However, as microprocessors are more widely used and their benefits recognized, engineers and designers will learn programming and begin to implement their systems with these devices.

Where are Microprocessors Used?

There are so many applications for microprocessors that it is difficult to classify and list them. However, to give you a glimpse at the many diverse uses for these devices, consider some of the applications where they are now being used.

1. Electronic Cash Registers
2. Electronic Scales
3. Electrical Appliance Controls
4. Automotive Controls
5. Traffic Signal Controllers
6. Machine Tool Controls
7. Programmable Calculators
8. Automatic Test Equipment
9. Data Communications Terminals
10. Process Controllers
11. Electronic Games
12. Data Collection

These are only a few of the many applications presently implemented with microprocessors. Just keep in mind that the microprocessor can be used in any other application where hard-wired standard logic systems are now used. In addition, microprocessors can also be used as the CPU in a small general purpose microcomputer or minicomputer.

A typical application for a microprocessor is illustrated in Figure 6-27. Here, the microprocesser is used in an electronic scale for a grocery market. The item to be weighed is placed on the scale. A transducer and analog-to-digital converter convert the weight into a binary word that is read into the CPU under program control. A clerk enters the price per pound via the keyboard. This too is read into the CPU. Then the CPU computes the price by multiplying the weight by the price per unit of weight. Then the total price is displayed on a 7-segment LED readout and printed on a ticket. All of this takes place under the control of the dedicated program stored in the ROM. Note the single 8-bit bi-directional data bus over which all data transfers take place.

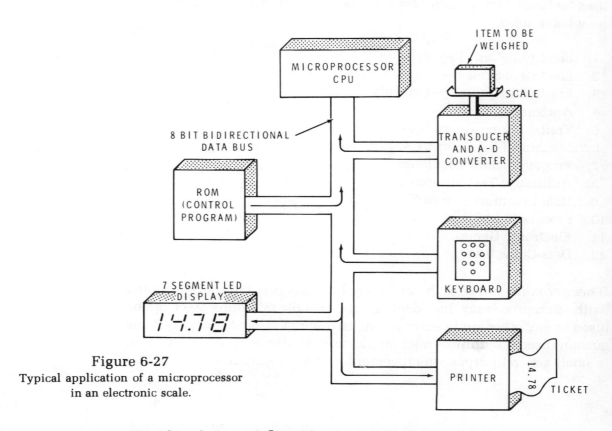

Figure 6-27
Typical application of a microprocessor
in an electronic scale.

Designing with Microprocessors

As indicated earlier, microprocessors can be used in two general ways. First, they can be used to replace minicomputers for dedicated control functions. Second, microprocessors can be used to replace standard hard-wired random logic systems. This section provides you with some guidelines to help you decide when and where a microprocessor should be used.

Microprocessors are generally much slower and less sophisticated than the typical minicomputer. But despite these limitations, microprocessors can often be used to replace minicomputers in some systems. The reason for this is that most minicomputers used in control systems are not used to their full capability. In a sense, they are a case of overkill. Many control systems used the minicomputer simply because of the ease with which the control can be changed by modifying the program. The significantly higher cost has been traded off for the convenience of system modification. In these applications, the microprocessor can usually handle the control functions as well as the minicomputer. A careful study must be made in such designs to see when a microprocessor can replace a minicomputer. There are many trade-offs to consider (speed, cost, etc.). Keep in mind that microprocessor development is in its infancy. Many technological improvements will be made over the years, causing the microprocessor to further approach the capabilities of today's minicomputer.

The microprocessor is a design alternative which should be considered in the early design stages of any digital system. The benefits of a microprocessor over standard hard-wired designs is significant in the larger, more sophisticated digital systems. As a general guideline, a microprocessor can be used beneficially if it will replace from thirty to fifty standard MSI and SSI TTL integrated circuits. If a preliminary design indicates that this many TTL integrated circuits must be used, a microprocessor should be considered. Unless the speed limitaton of the microprocessor is a factor, all of the benefits mentioned earlier will result by using the microprocessor.

Another way to equate a microprocessor design with the more conventional hard-wired logic design, is to compare the number of gates in a hard-wired design with the number of bits of memory required by a microprocessor system. It takes approximately 8 to 16 bits of memory in a microprocessor system to replace a single gate. Since most read only memories used to store the program for a microprocessor can contain as many as 16,384 bits, such a memory can replace from 1000 to 2000 gates. Depending upon the number of gates per SSI or MSI package, this can represent a replacement of hundreds of integrated circuit packages. A 16,384 (16K) bit ROM in a single 40-pin IC package, for example, can replace one hundred to four hundred 14, 16, or 24-pin SSI and MSI packages. This is a significant saving.

At this point, you may still have some doubts about the ability of a microprocessor to replace standard hard-wired logic functions. It may be difficult for you to imagine how a microprocessor can perform the functions you are so used to implementing with SSI and MSI packages. Therefore, we will consider all of the standard logic functions and illustrate how a microprocessor can perform them.

The microprocessor can readily perform all of the standard logical functions such as AND, OR, and Exclusive OR. It usually does this by executing the instructions designed for this purpose. Logical operations are generally performed on data stored in memory and in the accumulator register, with the result appearing in the accumulator. Suppose that you wanted to perform the NAND function on two 8-bit words. Using the instruction set in Table I, we could write the following program. Assume that the two words to be NANDed are stored in locations 8 and 9.

```
0   LDA   (8)
3   AND   (9)
6   CMP
7   HLT
```

The first instruction loads the first word into the accumulator. The second instruction performs the AND function with the word in the accumulator and the word in location 9. The result appears in the accumulator. Finally, this result is complemented to form the NAND function. This simple example illustrates the procedure you use to implement any Boolean function.

Arithmetic operations are also readily performed by a microprocessor. Special adders, subtractors and other arithmetic circuits are not required because all microprocessors can perform arithmetic operations through programming. Multiplication and division operations are carried out by subroutines. Even the higher math functions such as square root, trigonometric functions, and logarithms can be computed with subroutines. Many special algorithms have been developed for solving these higher mathematical functions with digital computers. To handle very large or very small numbers or to improve the accuracy of computation, multiple precision arithmetic subroutines are also available. Number size is limited by the number of bits in the basic computer data word. However, several computer words can be used to represent a quantity as large or as small as needed. Special programs can then be written to manipulate this data just as if it were represented by a single smaller word.

An example of a programmed arithmetic operation is shown in Figure 6-28. This flow chart illustrates the procedure for multiplying two positive numbers, A and B, by repeated addition. A is added B times to produce the product. The program to implement this algorithm is given below. The numbers to be multiplied are stored in locations 31 and 32. The product or answer is stored in locations 33.

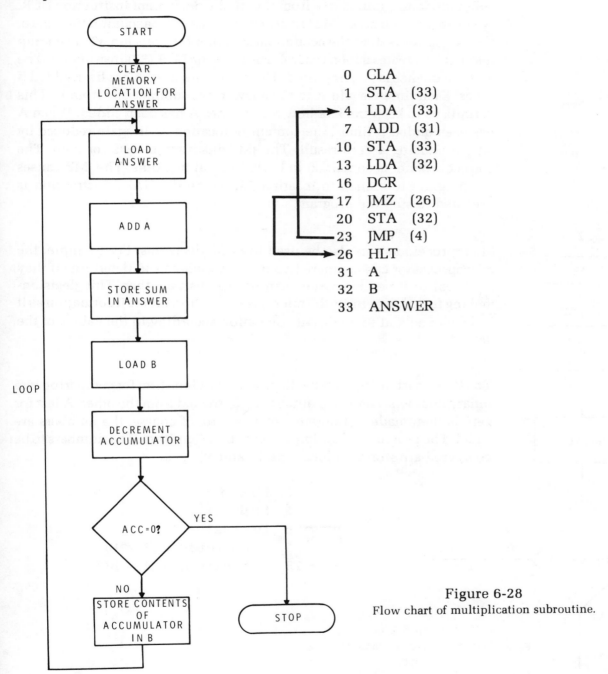

0	CLA	
1	STA	(33)
4	LDA	(33)
7	ADD	(31)
10	STA	(33)
13	LDA	(32)
16	DCR	
17	JMZ	(26)
20	STA	(32)
23	JMP	(4)
26	HLT	
31	A	
32	B	
33	ANSWER	

Figure 6-28
Flow chart of multiplication subroutine.

The first two instructions are used to clear the memory location where the ANSWER is to be stored. CLA resets the accumulator to zero and the STA instruction writes zeros in memory location 33. Next, the LDA instruction loads the contents of 33 (zero) into the accumulator. Then we add A to it with the ADD instruction. We then restore the partial product in location 33. We then load the content of location 32 (B) into the accumulator and subtract one from it with the decrement instruction DCR. We use a jump on zero (JMZ) instruction to see if the accumulator is zero. If it is not, we restore the accumulator content in location 32. The jump instruction creates a loop that returns us to the LDA (33) instruction. The entire sequence is then repeated. This continues until A has been added B times. Each time we add A to the answer, we subtract one from B. This permits us to keep track of how many times A has been added. When A has been added B times, the content of location 32 is again reduced by one, producing a zero result. The JMZ instruction tests for zero. The correct product is contained in location 33 at this time. The JMZ causes the program to branch to location 26, where the HALT instruction is executed to stop the program.

Microprocessors can also be used to make decisions. For example, the microprocessor can compare two binary numbers and determine if they are equal or if one is greater than or less than another. This decision-making function permits the microprocessor to evaluate information as it is developed and to modify its operation according to the values of the data.

The flow chart in Figure 6-29 illustrates one algorithm for comparing two binary numbers. Here, one number is subtracted from the other. A test for zero is then made. If the remainder is zero, of course, the numbers are equal. The program below implements this algorithm. The numbers to be compared are stored in locations 15 and 16.

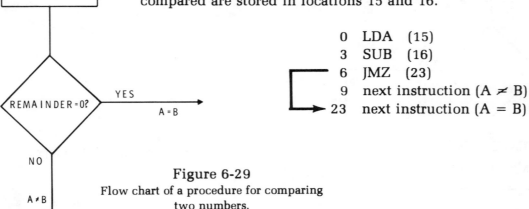

0	LDA	(15)
3	SUB	(16)
6	JMZ	(23)
9	next instruction (A $\neq$ B)
23	next instruction (A = B)

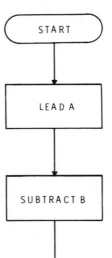

Figure 6-29
Flow chart of a procedure for comparing
two numbers.

If the numbers are equal, the program branches to location 23. If the numbers are not equal, the program continues in its normal, sequential manner.

Another common logic function that is readily implemented with a microprocessor is counting. The microprocessor can count external events or a frequency standard. External events are counted by applying them to the interrupt line on the microprocessor. As each event occurs, an interrupt is generated with the microprocessor. This causes the microprocessor to jump to a subroutine that will increment the accumulator register or add one to some memory location. Up or down counters are readily implemented with the increment and decrement accumulator instructions. Decision-making techniques can be used to detect when a specific count is reached or to count quantities larger than the computer word size permits. For example, with an 8-bit data word in a microprocessor, the maximum count that the accumulator can handle is 1111 1111 or 255. To count to higher values, a program can be written to indicate each time the counter overflows.

The program below illustrates a method of detecting a count of 153. The flow chart in Figure 6-30 shows the approach.

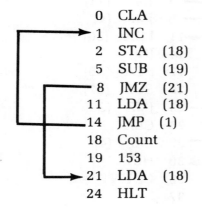

0	CLA	
1	INC	
2	STA	(18)
5	SUB	(19)
8	JMZ	(21)
11	LDA	(18)
14	JMP	(1)
18	Count	
19	153	
21	LDA	(18)
24	HLT	

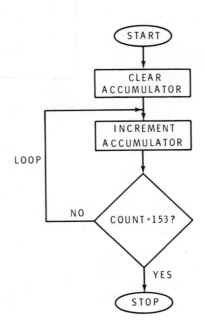

Figure 6-30
Flow chart showing a method of detecting a count of 153.

The first instruction clears the accumulator. The accumulator is then incremented by the INC instruction, and the count is stored in location 18. The count is then compared by subtracting 153 and testing for zero. If a non-zero result occurs, the count is retrieved with the LDA instruction and the program loops back to the increment instruction. This loop continues until a count of 153 is reached. When the JMZ instruction detects the zero condition, the program branches to location 21, where the count is loaded and the program halts.

To count to numbers higher than 255, the program below can be used. See the flow chart in Figure 6-31 for an explanation of the procedure. This program counts in multiples of 256. Note that the program has two loops. The inner loop determines when a count of 256 occurs, while the outer loop determines the number of times that the inner loop occurs. The total count then is the product of the number of times the inner loop occurs and the count in location 37, in this case 5. The program halts on a count of 5 × 256 or 1280.

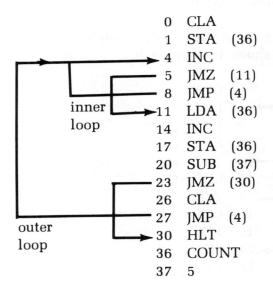

```
        0   CLA
        1   STA   (36)
        4   INC
        5   JMZ   (11)
        8   JMP   (4)
       11   LDA   (36)
       14   INC
       17   STA   (36)
       20   SUB   (37)
       23   JMZ   (30)
       26   CLA
       27   JMP   (4)
       30   HLT
       36   COUNT
       37   5
```

inner loop

outer loop

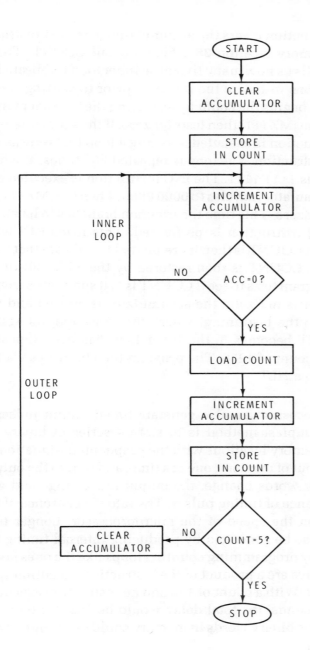

Figure 6-31
Flow chart illustrating a program to count
to 1280 (5 × 256).

The first instruction clears the accumulator. The next instruction writes zero into memory location 36 which we call COUNT. The content of location 36 tells us how many times the inner loop is repeated. These first two instructions initialize the circuitry prior to starting the count. The program then begins the count by executing the increment instruction in location 4. The JMZ (11) then tests for zero. If the accumulator is not zero, the JMP instruction is executed, creating a loop that returns the program to the INC instruction. The loop is repeated 255 times, at which time the accumulator is 1111 1111. The INC instruction is executed a 256th time and the accumulator recycles to 0000 0000. Then the JMZ (11) instruction again tests for zero. This time the program branches to location 11 where the LDA (36) instruction is performed. This loads COUNT (which is initially zero). COUNT is then incremented to indicate that a count of 256 has occurred. COUNT is then restored by the STA (36) in location 17. Next, the program tests to see if COUNT is 5. It subtracts 5 from COUNT. If the remainder is not zero, the accumulator is cleared and the program loops back to the beginning, where the inner loop is again repeated. When COUNT becomes 5, the inner loop has been repeated 5 times, indicating a count of 1280. The program then branches via the JMZ (30) instruction to a HLT.

The microprocessor can also generate timed output pulses, in several ways. The simplest method is to store a series of binary numbers in sequential memory locations with the proper bit designations. These can then be read out of memory, one at a time, and sent to the output data bus. As the binary words change, the output bits change and generate any desired sequence of timing pulses. The rate of occurence of these pulses depends upon the speed of the microprocessor. Longer timed output pulses can also be generated by producing internal timing delays. This can be done by programming counting loops like the ones just illustrated. The time delays are a product of the instruction execution speed and the desired count. With a count of 153 and an instruction execution speed of 12.5 microseconds, the total delay would be $153 \times 12.5 = 1912.5$ microseconds. The binary words in memory could be outputted every 1.9125 milliseconds.

Self-Review Questions

41. The most popular and widely used microprocessor is a _____
    _____ circuit.

42. Which of the following best describes a microprocessor?

    A. A general purpose digital computer.
    B. A special purpose digital computer.

43. What are the two general applications for microprocessors?

    1. _____
    2. _____

44. List the four benefits of using microprocessors to replace hard-wired
    logic.

    1. _____
    2. _____
    3. _____
    4. _____

45. Write a program showing how you would multiply a number (X) in
    location 22 by 8. Use the instruction set in Table I. Start your
    program in location 0. (Hint: A shift-left operation multiplies by 2.)

Self-Review Answers

41. The most popular and widely used microprocessor is a MOS LSI circuit.

42. B — A microprocessor is best described as a special purpose digital computer.

43. The two general applications for microprocessors are:

   1. To replace minicomputers.
   2. To replace hard-wired logic.

44. The four benefits of using microprocessors to replace hard-wired logic are:

   1. Reduced development time and cost.
   2. Reduced manufacturing time and cost.
   3. Enhanced product capability.
   4. Improved reliability.

45. The flow chart in Figure 6-32 shows one method of multiplying the content of location 22 (X) by 8. The program is given below.

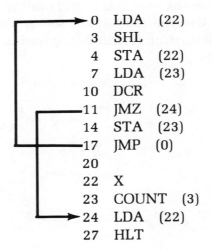

```
 0   LDA   (22)
 3   SHL
 4   STA   (22)
 7   LDA   (23)
10   DCR
11   JMZ   (24)
14   STA   (23)
17   JMP   (0)
20
22   X
23   COUNT  (3)
24   LDA   (22)
27   HLT
```

Each time the number X is shifted left, it is effectively multiplied by 2. Three shifts produce multiplication by 8. A counter and decision-making loop determine when three shifts occur.

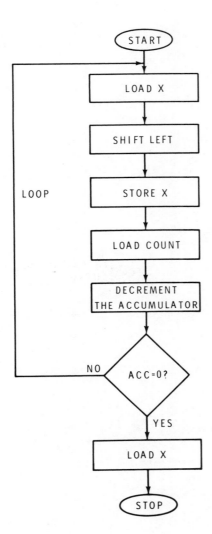

Figure 6-32
Flow Chart illustrating the procedure for
multiplying a number by 8 by shifting.